L...
i...
M... ms

Ma... relle

3

D

F
S

Theory, Tests and Applications

Springer-Verlag
Berlin Heidelberg New York London Paris Tokyo

**Author**

Dr. D. van der Wijst
Research Institute for Small Business
P.O.Box 7001
NL-2701 AA Zoetermeer, The Netherlands



ISBN 3-540-50574-1 Springer-Verlag Berlin Heidelberg New York
ISBN 0-387-50574-1 Springer-Verlag New York Berlin Heidelberg



Printing and binding: Druckhaus Beltz, Hemsbach/Bergstr.
2847/3140-543210

CONTENTS

## List of tables

# PART I
# introduction

# 1 AIM AND OUTLINE OF THE STUDY

## 1.1 Purpose and motivation

The subject of this study is the small firm's choice of the level and composition of its investments and of the sources from which the funds for these investments are attracted, or in short: the small firm's financial structure. This subject is studied from several points of view. Small business characteristics are described and the literature is surveyed to collect hypotheses regarding the determinants of financial structure[1]. These hypotheses are tested, as far as possible, in empirical analyses and their applicability is investigated. None of the topics in this range is particularly emphasized –the study is broadwise rather than specialized– and the various points of view are interrelated, e.g. applicability plays a role in the empirical analyses.

This book is the result of a research project into small business finance, which has been carried out by the Research Institute for Small and Medium-Sized Business in The Netherlands (henceforth: EIM, the Dutch acronym) in co-operation with the Erasmus University in Rotterdam (EUR); The study has a Dutch background. The project is part of the fundamental research programme of the EIM, which was initiated in 1973 as a joint research effort of the EIM and the Econometric Institute of the EUR[2]. In this programme, various aspects of the small firm's operations are analysed; earlier elements of the programme include analyses of labour productivity, floorspace productivity, price setting, etc. (Nooteboom, 1980, Thurik, 1984).

The present investigation of small business finance is actuated by the empirical fact that large differences in financial structure both between and within small business industries can be observed, while little is known about the causes or even the extent of this phenomenon. The research project addresses this subject with the purpose to develop theoretically and empirically well founded models which explain differences in financial structure between and within groups (or: industries) of small and medium-sized enterprises (SME's). Although scientific interest would by itself be a valid reason to initiate such a research project, the inquiry

-------

[1]*Note that "financial structure", as it is defined here, refers to both assets and liabilities/equity. The same term is sometimes used in a more narrow sense, referring exclusively to liabilities/equity* (*e.g.* Haley *and* Schall, 1979), *although the latter is more often denoted by the term "capital structure".*

[2]*See* Thurik (1984) *for a survey of the history of this programme.*

into this socially and economically important subject is also induced by a number of practical considerations. The common denominator of these considerations is a lack of information regarding almost every aspect of the way SME's are financed. Consequently, a proper assessment of the financial structure of small businesses on all levels of aggregation is thwarted and this, in turn, seriously hampers a wide range of decisions that are more or less regularly made by small businessmen, bank employees, investors and even government[3].

The first aspect that should be mentioned in this context is the shortage of data describing the actual financial structure of SME's. A major reason for this shortage is the fact that for the majority of small firms publication of the annual report is not mandatory and, hence, simply not done. This means that data collection is not a matter of picking up financial statements and standardizing them, but a matter of laborious field work. The resulting data-scarcity can be illustrated by quoting a report, issued almost a decade ago by the Council for Small and Medium-Sized Business[4], which states that, with a solitary exception, over the period 1965–1971 balance sheet data of not a single small business industry are available (Raad voor het MKB, 1979, p.8). Since then, the situation has improved, among other reasons because the EIM started publishing balance sheet data of SME's in the retail trade[5], but even today large parts of the small scale manufacturing and service industries are virtually terra incognita. Although data collection is not a primary objective of the research project, it enhances the accessibility of already available data by providing a catalogue and it describes some data that have not, or only partly, been published before.

The second aspect of small business finance that is still incompletely mapped regards the various possible determinants of financial structure. Determinants and potential determinants of the firm's financial structure are discussed in a range of publications, which are quite diverse in background and character. On the one hand, some broad categories of debt ratio determinants are described in the theoretical studies of business finance, that have been brought out since the publication of Modigliani and Miller's (1958, 1963) seminal papers. However, the theoretical developments as well as their empirical implications are primarily discussed in the context of large

-------

[3]*For instance, in The Netherlands government gives security for certain bank loans to SME's.*

[4]*An advisory council of the Minister of Economic Affairs.*

[5]*See, for instance, the publications "Balanspositie* 19..", *issued by the EIM with respect to the years* 1984, 1985 *and* 1986.

firms listed on stock exchanges and the relevance for small business is hardly, if at all, expounded[6]. On the other hand, the every day practice of small business finance is depicted in a multitude of advisory and descriptive studies, that have been brought out parallel to, but mainly independent of, the developments in the theory. This practical literature[7], however, presents conclusions that are often entangled and not always flawless, while the absence of a relation with the theory is also a drawback. So, on the whole, no clear picture of the determinants of small business finance emerges as yet from the collective literature, although certain key elements are unmistakable. A logical first step would be to make an inventory of financial structure determinants proposed in both the theoretical and the practical literature to see how they are related to each other. The research programme comprises an effort to do so, and by subsequently analysing the differences and similarities between the determinants from both sources, we may provide a firm footing for empirical analyses and possibly throw some light on the complex capital structure puzzle.

The third aspect of small business finance that needs considerable elaboration concerns the empirical evidence supporting the various suggested determinants. The theoretical contributions in particular are scarcely tested on their relevance for the small business sector, but some "practical" determinants also deserve little more than the qualification "assumption". Empirical tests of the hypotheses, obtained by reformulating the inventory of determinants, are a major element of the research project, although the data may, of course, restrict the possibilities to test certain hypotheses.

Finally, the fourth aspect of small business finance that is still far from being completely mapped pertains to the application of research outcomes. Once data are gathered, a list of possible determinants is drawn up and their relevance is empirically tested, the question remains how the results of all this can be used for small firm financial management, for interfirm comparisons and credit evaluations, etc. In the current practice, ratio analysis is nearly the only instrument used for these purposes, but this simple technique clearly falls short if complex evaluation schemes, involving multiple determinants, are to be applied. In the research project, attention is paid to the application of the results and procedures are outlined to convert complex estimation results into simple management tools.

-------

[6] *Even the so-called "small firm effect" in the theoretical literature refers to firms listed on stock exchanges.*

[7] *The distinction between theoretical and practical literature is, of course, somewhat artificial but useful to accentuate the differences.*

## 1.2 Scope and methodology

The aspects of small business finance that are mentioned in the previous section are the principal elements in the research project. Together, they cover a wide range of subjects and in order to keep the research project amenable, its scope obviously has to be limited in several respects. This means that each of the subjects discussed in this study could be elaborated in more detail in a separate, more specialized study, but that is not the approach followed here. The most important limitations regarding each of the four elements (data, inventory of determinants, empirical analyses and application) are discussed in this paragraph, along with the strategic choices that are made in the matter of research methodology.

As regards the data, only existing sources are tapped, i.e. no new data are collected. The data that became available in this way are not always recent, they describe financial structure in accounting terms rather than in market values and the variables best suited to test certain hypotheses may not be present, to mention but a few of the drawbacks. However, many of the limitations are inherent to empirical research in a small business environment, so they would not be avoided by collecting new data, and drawing on existing sources enables the use of a considerable amount of data. Two time-series (covering 10 resp. 20 years) of industry averages in retailing are analysed, as are cross-section data of more than a dozen samples of individual SME's.

The inventory of financial structure determinants is primarily made on the basis of the literature. Suggestions from the practice of small business finance are exclusively taken from the literature, no empirical research into the subject is carried out. The theory, on the other hand, is not only studied by way of the literature, but also by constructing some models of capital structure choice. In these models, major recent developments in the theory of optimal capital structure are systematically restated in a mathematical form, based on a uniform framework of definitions and derivations. The purpose of this procedure is to eliminate distracting differences in methodology between the various contributions, thus facilitating a clear overview and a proper assessment of the developments. The framework adopted for this restatement is the extended Modigliani–Miller theory, modelled within the so-called mean–variance approach under a set of fairly restrictive assumptions. This particular framework is chosen because it is comparatively simple and because, over the years, it has been used to model a considerable part of the contributions to the theory. These characteristics fit in well with the purpose of the theoretical exercises, which is to inventory financial structure determinants in the literature rather than to advance the theory itself. For the latter purpose, more recent theories as the Black and Scholes (1973) Option Pricing Theory and the Ross (1976) Arbitrage Pricing Theory, on which today's professional interest seems to centre, would be a more obvious framework to choose. However, the application of these theories to the problem of the firms optimal capital structure is complex and still in the early stages of development (see e.g. Kemna, 1987). In the future, these theories may provide a firmer footing for subsequent (empirical) analyses. So the model construction efforts largely follow the Modigliani-Miller theory. Besides providing a uniformly derived overview, this may also throw some light on the robustness of the various theoretical contributions. Modelling these contributions within a single framework should, in principle, produce a tree-structured family of interrelated models, obtained by subsequently relaxing assumptions or introducing new phenomena. However, in view of the general tendency to publish only well-behaved models, this may not be the case, especially not as regards the "intermediate" models, i.e. non-published variants of well-known models, and this area is still far from being completely mapped. Establishing the length to which the beaten track of the MM-theory is indeed practicable could in itself be a contribution to the theory .

The inventory of financial structure determinants can be reformulated as a list of testable hypotheses, which are the subject of the empirical analyses. A broad variety of methods and techniques is available to test the hypotheses and the choice between them is based on the following considerations. First of all, the technique should allow an adequate expression of the arguments for a determinant's influence, e.g. the use of arbitrary groupings to express a continuous influence should be avoided as far as possible[8]. Secondly, the technique should have some connections with the practice of financial analyses in small business. Since the latter usually involves ratio analysis, the technique should produce results that, in extreme simplification, can be considered equivalents of commonly used ratios. Thirdly, and partly as a result of the foregoing, the technique should facilitate an easy application of the results, both technically (e.g. not require an excessive amount of calculations) and conceptually (i.e. lend itself to popularization). From the above it will be clear, that the use of

-------

[8]*Examples of such groupings can be found in many applications of multiple discriminant analysis in the field of finance.*

advanced econometric techniques is not pursued as an aim in itself; from an econometric point of view, the empirical analyses can be described as exploratory. However, techniques are chosen to adequately reflect the purpose of the project in view of the available data. These considerations lead to the use of statistical techniques (primarily least squares estimation of non linear relations) that are by now familiar in the academic world, but that are still virtually unknown to financial analysts in small business. This accentuates the need to establish a link with practice.

As regards the application of the results, the scope of the research project is limited to an illustration of the possibilities, a ready to hand product is not aimed at. The reason for this limitation is twofold. Firstly, any application should be based on recent data, whereas the research project uses data that, although well suited for hypothesis testing, are mainly not recent enough for application of the results. Secondly, the research project aims at developing evaluation schemes (models) for the financial structure of SME's that involve multiple determinants and that, therefore, may be rather complex. In order to make these results accessible to the small business community, they have to be distributed through "intelligent" media, i.e. be embedded in software that translates the determinants relevant for an individual firm into a "yardstick financial structure" for this particular firm. The development of this software is beyond the scope of the research programme.

## 1.3 Organization of the study

This study consists of four parts. The present, introductory part is concluded with chapter 2, in which we discuss the characteristics of small business, specifically the financial characteristics. The many definitions of small business, that appear to be in circulation, are briefly reviewed in this chapter, as is the general importance of small business for all western economies. A large number of financial characteristics of small business, collected from the literature, is summarized but no clear pattern seems to emerge from the multitude of suggestions.

The second part of this study, comprising four chapters, contains an inventory of financial structure determinants. In chapter 3 we give a verbal account of the developments in the theory of optimal capital structure over the past three decades. Some broad categories of debt ratio determinants, which are the results of these developments, are discussed along with their empirical implications and the variables that are commonly used to approximate their effects in empirical analyses. In chapter 4 theoretical aspects of the financing decision are studied by constructing some models of capital structure choice. It will be show that, even within the simple and rather restrictive framework of assumptions adopted in this study, the robustness of the various theoretical models leaves much to be desired. Many (published) models appear to be sensitive for small, and by themselves trivial, changes in their specification so that the prospects of further extending these models with more realistic assumptions are at least questionable. Chapter 5 is devoted to the financial structure determinants that are suggested by the literature on the practice of (small) business finance. The subjects addressed and the methods used in this literature vary widely, but some categories of financial structure determinants can be distinguished and in each of them, a large number of influences on the firm's financial behaviour is discussed. We will argue that a number of them are motivated in a somewhat questionable matter. The similarities and differences between the determinants of financial structure presented in both the theoretical and the practical literature are discussed in chapter 6. We will conclude that both approaches differ more in method than in content and that they supplement each other both in their description of the small firm's financial behaviour and in their contributions to the empirical analyses.

Empirical analyses are the subject of the third part of this study. Two series of empirical analyses, using data on different levels of aggregation, are reported on. Data of individual firms within more or less narrowly defined industries are analysed in chapter 7. We use fourteen samples in this study, most of which refer to the retailing trade but samples from manufacturing industries are also included. Almost all samples contain detailed information of the firms involved, so that we are able to test 15 different hypotheses, albeit in different subsets for different samples. In chapter 8 we analyse a Dutch and a German database of industry averages in the retail trade. Both financial structure (in terms of money amounts) and changes in financial structure (first differences between money amounts) are analysed.

Application of research outcomes, such as those obtained in this study, are a somewhat neglected subject in the literature. In the fourth and final part of this book, which contains two chapters devoted to applications and conclusions, we explore the possibilities of application. In chapter 9 we describe some obstacles to applications, both on the theoretical side and on the practical side. Subsequently, we illustrate the possibilities to apply the research outcomes of chapter 8 to the widely used practice of interfirm comparisons and we demonstrate the potential value of such applications. We end our study with chapter 10, where we summarize the contents and the main conclusions of this study.

## 2.1 Introduction

The appreciation of small-scale economic activity has undergone a remarkable change in recent years. In the sixties and early seventies, economic progress and technical development were expected to come from large firms, which were, according to some authors, "an almost perfect instrument for inducing technical change" (K. Galbraith, quoted by Williamson, 1975). Small business, on the other hand, was usually associated with peripheral or backward areas of the economy, characterized by negative qualities as inefficient use of resources and low quality employment[1]. Things have, however, changed since then. The economic recession and mass unemployment of the late seventies-early eighties shook confidence in the traditional pillars of the economy and turned the scales in favour of small business. The pivotal publication in the process of making "small" almost identical to "beautiful" was perhaps "The job generation process" by Birch (1981). After that, the importance of small business has been underlined by many authors in often colourful but not always factual terms[2], while virtually no negative opinions of small business have since been expressed in the literature. Among the more popular qualities attributed to small business is that of a breeding-place for new products, new technologies and new industries, with software and micro-computers as favourite examples (Adams, 1982, Struck and Mandell, 1983). However, this contention usually cannot stand the test of serious research (see e.g. Freeman, 1982, Binks and Jennings, 1986), as is the case with other currently popular beliefs regarding the role of small business in the economic and social revitalization of society. So, despite the recent reappraisal of small-scale economic activity that originates from the rediscovery of the small firm as a material source of job creation, it still seems necessary to restate some facts about the importance of the small business sector as an introduction to any serious study in this field. This is particularly true for a study of financial structure in small business,

-------

[1]*Of course, there are always authors who remain unaffected by the economic fashion. For instance, in* 1968 *Robinson, contrary to the conventional wisdom of the day, promoted small business as the economic manifestation of the Great American Dream* (*Robinson*, 1968).

[2]*To give but one example: J. Kenworthy and M.A. Wilmer presented a paper at the* 9*th European Conference of Operations Researchers in Paris* 1988 *under the title "In the ocean of commerce, small firms are the plankton".*

since most writings on the theory of finance and its empirical implications have alluded to large, listed firms. However, the question to be answered prior to any such treatise is, of course, what is meant by "small business", while some attention should also be paid to the, real or alleged, peculiarities of small business finance. These issues are addressed below.

## 2.2 Some definitions

In view of the great diversity in character of the business community, defining "small business" is not a straightforward operation and many definitions have been suggested. In The Netherlands, small and medium-sized business comprises all privately held firms, except those in the agricultural- and fishing-industry, with 1 through 9, respectively 10 through 99 persons employed[3] (Braaksma et al., 1987). In Sweden, small and medium-sized business is generally defined as firms with less than 200 employees, while in England the same maximum number of employees is used to define only small firms. SME's[4] in Germany may employ up to 500 people, and this definition also seems to be used by the European Economic Community authorities, with various sub-divisions depending on the type of industry. In Japan, which has one of the largest small business sectors, the term "small" refers to firms employing up to 20 people, and the upper limit for small and medium-sized in Belgium varies from 40 to 70 employees (Claessens, 1982). Of the many definitions that have been put forward in the United States, the one that is formulated by the Small Business Administration is probably the most widely used. According to this definition, a small firm is independently owned and operated and is nondominant in its field of operations. Further specifications include an average employment of less than 250 for manufacturing companies, sales sizes below $5 to $15 million for wholesale firms and sales or receipts below $1 to $5 million for retailing and service firms (see e.g. Walker and Petty, 1978a, p. 3). Although this enumeration is far from being complete, it is amply sufficient to illustrate the lack of uniformity[5]. The reason for this profusion of definitions is, of course, the

-------

[3] *Until a few years ago, the upper boundary was set at 50 employees.*

[4] *The terms "firm", "business" and "enterprise" are used interchangeably, as may have become clear by now.*

[5] *Moreover, it should be noted that in most countries different definitions are used for different purposes, so that various definitions may occur.*

fact that industries vary widely in average size, so that it is almost inevitable to put greatly different quantitative size limits on them[6]. Depending on the type of industry one has in mind or that is prevailing in a certain country, a different definition is likely to be used: a small steel producer or car–manufacturer is indeed much larger than a large plumber's, hairdresser's or butcher's shop.

## 2.3 Importance

Irrespective of the precise definition used, there is no doubt that the small business sector is an important part of the economy in all western countries. Table 2.1 pictures the structure of the US–economy by employment size of business and table 2.2 summarizes some key data on the importance of SME's in The Netherlands. These tables show that by numerical count small business dominates the economic scene: by far the majority of firms is classified as small. In terms of employment, receipts or value added the share of small business is lower but still impressive, as also becomes evident from other publications. Bannock (1986), for instance, presents statistics on the share of small business in manufacturing employment in 13 countries. The figures, referring to the mid–seventies, range from well over 60% in Japan and Switzerland to below 30% in the U.K. The yearly report of the president on

Table 2.1 The structure of the U.S. economy (percent)

| | | Employment size of business | | |
|---|---|---|---|---|
| | total | under 100 | 100–499 | 500 or more |
| Firms | 100 | 98.5 | 1.3 | 0.2 |
| Employees | 100 | 41 | 12 | 47 |
| Receipts | 100 | 33 | 12 | 55 |
| Assets | 100 | 22 | 13 | 65 |
| Net income | 100 | 26.7 | 8.2 | 65.1 |

*Note Table* 2.1 :
*Source : The state of small business : a report of the president*, 1986, (*Washington, U.S. Government printing office*).

---

[6] *Robinson* (1986) *mentions that the smallest "small" size limit may be one–hundredth of the largest "small" size limit.*

the state of small business in the USA contains a multitude of statistics and, occasionally, thorough descriptions of the role of small businesses in the national economy as demand shock absorbers, as agents in resource realignment, as cultivators of labour resources, as suppliers of specialized goods and services, etc., etc. A host of other figures could be mentioned; however, the above information probably suffices to underscore the general economic importance of small business.

Since data regarding the financing of small firms are rather scarce in public sources, the importance of small business is harder to depict in financial terms than in general economic terms. Table 2.1 shows that in the USA small firms hold 22% of all assets and account for 33% of receipts and 26,7% of net income. The yearly reports on the state of small business in the USA (see e.g. note table 2.1) contain many more of these facts: finance companies (or "asset based lenders") extended 50% of their \$143.6 billion ($10^9$) outstanding receivables to small businesses, 44.6% of all new direct finance leasing volume went to lessees with fewer than 100 employees, etc. The general picture, that emerges from the myriad of little facts, is that small business accounts for roughly 25% to 50% of the funds involved in business finance.

Table 2.2 Data of small and medium-sized versus large privately owned firms in The Netherlands

| | Year | Small and medium-sized | Large (>100 empl.) |
|---|---|---|---|
| Number of firms | 1986 | 413.900 | 5.100 |
| Employment (full time equivalents) | 1986 | 1.609.400 | 1.192.800 |
| Value added (fl.million) | 1985 | 101.600 | 102.800 |
| % exporting firms | 1985 | 7 | 58 |
| Stock of tangible fixed assets (fl.million)* | 1985 | 89.000 | 99.000 |
| Investments in tangible fixed assets (fl.million) | 1981 | 11.370 | 11.010 |
| | 1983 | 10.300 | 13.250 |
| | 1985 | 14.260 | 17.420 |

*Notes Table* 2.2 :
- *source* : *Braaksma et al.*, 1987;
- *privately owned firms include neither regulated firms as hospitals and public transportation companies nor firms in primary industries (mining, fishing and agriculture)*;
- * *banks and insurance companies not included.*

In The Netherlands, some tentative information on the financial importance of small business recently became available from a government sponsored survey (Braaksma, et al., 1987), some of which is included in table 2.2. These data indicate that the *f*188 million ($10^9$) worth of tangible fixed assets, held by privately owned firms in 1985, are more or less equally distributed over SME's and large firms, and the same can be said as regards the yearly investments in tangible fixed assets. These figures can be compared with similar data of firms listed on the Amsterdam Stock Exchange, as the Dutch Central Bureau of Statistics publishes a yearly summary of their financial statements[7]. The 1985 publication comprises 118 (large) firms, which in 1983 invested *f*19.9 million in fixed assets and presented roughly *f*125 million worth of tangible fixed assets on their balance sheets. However, the five international concerns[8] listed in Amsterdam, which hold a large part of their assets abroad, account for three-fourths of these amounts. Taking that into consideration it seems safe to say that in terms of assets and investments the small business sector in The Netherlands is at least as large as the domestic part of the listed firms, but probably larger[9]. By inference, approximately the same can be said of the other items on the balance sheet, so that in terms of debt and equity SME's are also a match for the listed firms. Although the financial importance of the small business sector may be different in other countries, the American and Dutch examples shown here demonstrate, that the small firms" claims on productive assets and capital are substantial enough to merit more study than is presently devoted to them.

## 2.4 Financial characteristics

### 2.4.1 *Introduction*

Size related characteristics of business finance are not a standard element of textbooks on finance. In the early editions of the well-known textbooks by

-------

[7]*Centraal Bureau voor de Statistiek*, 1985, *Statistiek van balans en resultatenrekening beurs-N.V.'s* 1983. *Banks and insurance companies are not included in this publication.*

[8]*These are Royal Dutch Petroleum, Philips, Unilever, Akzo and Hoogovens.*

[9]*Other estimations of the financing volume in The Netherlands are also in circulation. For instance, Eijgenhuijsen et al.* (1987) *present a tentative calculation of the total amount of equity capital in The Netherlands. Including the primary industries and banks and insurance companies, they arrive at a grand total of f*360 *million* ($10^9$) *worth of equity, but no subdivisions are provided.*

Weston and Brigham (1972), Haley and Schall (1979) and Copeland and Weston (1979) no separate section is devoted to this subject, while Neave and Wiginton (1981) spend two puny sections on the matter[10]. However, textbooks seldom tell the whole story and although their lack of interest indicates that "size" is not a major issue in finance, a large number of publications on the subject can be found. The topics discussed in these publications are rather diverse in character. The so–called "small firm effect" gained considerable attention, albeit not as a result of a careful examination of the characteristics of small business finance, but as a result of an empirical "anomaly" in stock pricing models. In other publications, differences between small and large firms are empirically studied outside the context of specific stock pricing models. Still other publications refer to size related aspects of the organization of financial markets and institutions, or merely use size as one of many explanatory variables in a study. Furthermore, size effects are studied in business populations varying from SME's to Fortune's top 500 firms. Clearly, not all of these populations and phenomena are within the scope of this study, but they are nonetheless considered here because they may contribute to a better understanding of the influence of size on the financial behaviour of the firm. To structure this rather incoherent collection of literature, elements from the various publications are grouped around three main points, to be discussed in the following subsections. First of all, empirical studies that are not based on a specific stock pricing model are briefly reviewed in the next subsection. The small firm effect, extended into a more general discussion of the accessibility of stock exchanges for small firms, is discussed in section 2.4.3, followed by a tentative description of the small firm's capital markets in section 2.4.4.

### 2.4.2 *Some empirical studies*

Size influences on the financial behaviour of firms are reported on in two types of empirical studies, the small firm effect in stock pricing models not included. On the one hand, a number of more or less descriptive studies is devoted to this subject, while on the other hand size is used as an explanatory variable for the choice between debt and equity in security issues. The descriptive studies typically contrast financial data of two of more size groups and provide an enumeration of possible causes of differences

-------

[10] *Usually, the later editions have responded to the growing attention for the small firm and include a discussion of small business finance.*

between the groups. The analyses of security issues focus on the characteristics, including size, of firms that issue debt respectively equity. Table 2.3 summarizes the empirical findings of a dozen studies of both types, covering different countries and various time periods. As table

Table 2.3 Summary of financial differences between small and large firms

| Publication/ Author(s) | Data from country/period | Major findings |
|---|---|---|
| The State of Small Business (1986) | U.S.A. 1977–80 | there are as many industries in which the debt ratio increases,as there are industries in which the ratio diminishes with increasing business size |
| Huisman & Boon (1984) | The Netherlands 1983 | smaller firms use <u>less</u> debt, on average, than do larger firms |
| Marsh (1982) | U.K. <u>issues</u> 1959–74 | smaller companies are <u>less</u> likely to issue debt |
| Tamari (1980) | UK,USA,Japan, Israel, France 1964–73 | small firms have <u>average</u> (U.S.A., Japan), <u>above average</u> (U.K.,Israel) or <u>below average</u> (Fr.) debt ratios |
| Ferri & Jones (1979) | U.S.A. 1969–76 | small size is <u>negatively</u> related to the use of debt capital but not in a straightforward manner |
| Walker & Petty (1978b) | U.S.A. ....–71 | small firms have <u>more debt oriented</u> capital structures, low dividend payout ratios, less liquidity and are better at generating earnings |
| Scott & Martin (1975) | U.S.A. 1967–72 | small firms are generally associated with <u>small debt ratios</u> (low leverage) |
| Taub (1975) | U.S.A. <u>issues</u> 1960–64 | small size has a <u>negative</u> impact on the probability that debt is issued |
| Toy et al.(1974) Remmers et al. (1974) | Norway,U.S.A., France, Japan, The Netherlands 1966–72 | size does <u>not</u> appear to be a determinant of debt ratios |
| Baxter & Cragg (1970) | U.S.A. <u>issues</u> 1950–65 | the smaller the company the <u>less</u> likely it is to issue bonds rather than stock |
| Brigham (1967) | U.S.A. 1965 | small firms tend to use debt <u>more extensively</u> than do larger firms |

2.3 shows, there is a striking lack of agreement as to the effect of size on financial structure in these studies: a small size can be associated with high as well as low as well as average debt ratios. Only the three studies of security issues are unanimous as regards the direction of the size effect: a small size has a negative impact on the probability that debt is issued. Although there is no ready explanation for these seemingly random size effects, it should be noted that the results in table 2.3 refer to different industries and quite different size groups. Hence, the possibility of consistent size effects within certain industries and size ranges cannot be excluded on the basis of these results. However, the empirical findings in table 2.3, which are deliberately presented on a high level of aggregation, illustrate that there is no such thing as a uniform size effect across all industries and size ranges.

Enumerations of the possible causes of differences between size groups show an equally ambiguous picture as the results themselves. A broad variety of causes is suggested, the more relevant of which refer to the small firm's limited access to capital markets; these causes are discussed in the next subsection. Other potentially relevant causes include the small firm's limited bargaining power (Sprenger et al., 1982) and limited opportunities to shop for credit (Brigham, 1967). However, a surprisingly large number of suggestions do not refer to size as such. Scrutinizing the publications in table 3 and some closely related literature, produces the following list of non-size effects that are used to argue for size-related financial differences:

- industry effects: SME's tend to be clustered in highly fragmented industries as retailing and wholesaling (Welsh and White, 1981);
- collateral effects: SME's are frequently denied credit because they lack collateral (Sprenger et al., 1982) or have less resources to fall back on (Taub, 1975);
- risk effects: SME's tend to have more operating risks (Viscione, 1986), the bankruptcy rate of SME's is larger than that of large firms (Tamari, 1980) and SME's are less diversified (Ferri and Jones, 1979);
- volatility effects: SME's are frequently subject to seasonal variations in sales (Welsh and White, 1981) and have much greater fluctuations in the profit rate over time (Tamari, 1980);
- profitability effects: smaller (manufacturing) firms are less profitable than larger ones (Brigham, 1967) or, on the other hand, many small firms do earn profits higher than those of large firms.

Although it may be true that some, or even all, of these effects make themselves felt more strongly in small than in large business, they are neither inherently size related nor by any means restricted to small business. Hence, they are only circumstantial causes of size related financial differences.

### 2.4.3 *The small firm effect*

The small firm effect, which attracted considerable attention in the recent literature, refers to an empirical anomaly in stock pricing models (more specifically the CAPM). When the relation between return and total market value of listed stocks is analysed, small firms[11] appear to have higher risk adjusted returns, on average, than large firms and these returns are "abnormal" in the context of the CAPM. The discovery of this empirical anomaly gave rise to a broad variety of possible explanations of this phenomenon, which became known as the small firm effect. Size related transaction costs are an obvious candidate, as Roll (1981) pointed out, but he also mentions the possibility that the pricing model underestimates the risk of small firms with infrequently traded shares. Alternatively, James and Edminster (1983) suggest that investors may demand higher returns on infrequently traded shares to compensate for their limited marketability. Moreover, investors in small firms may also claim compensation for a lack of control over the firms, specifically as regards the possibilities to protect themselves against involuntary mergers and take-overs (see Van den Bergh et al., 1984). Still others, like Barry and Brown (1984), base their explanation of the small firm effect on information arguments. The high returns realized on investments in small firms may contain a compensation for the higher costs of gathering the necessary information on small firms, or a compensation for the additional risk if this information, or a part of it, is missing.

Despite the large number of suggestions made, none of the proposed arguments seems to provide a theoretically and empirically satisfactory explanation of the small firm effect[12]. However, the suggested arguments themselves are of value. By describing size related aspects of the way financial markets operate, they provide an overview of potential size related determinants of financial structure. These determinants are relevant to a

-------

[11] *Note that these small firms are probably large by any small business definition.*

[12] *See Van den Bergh et al.* (1984) *and Dorsman et al.* (1987) *for a comprehensive review of the literature and empirical evidence regarding the Dutch stock market.*

wider range of problems than the somewhat restricted anomaly in stock returns. Moreover, they provide a direct link with the small business literature, as becomes clear from the following. Obviously, the arguments for the small firm effect lose their relevance at the upper end of the size range: transaction and information costs become negligible, the claims to compensate for limited marketability and control vanish, etc. What remains is the perfect capital market for which financial economists have such a pronounced bias. Alternatively, if the size range is extended at the lower end, the effects are also clear: transaction and information cost become prohibitively high, the lack of marketability and control increase beyond compensation, etc. In short, the capital market becomes inaccessible for firms below a certain size. The conclusion, that the very size of small businesses prohibits access to large, nation–wide capital markets as stock exchanges is extensively referred to in the small business literature. The arguments for this inaccessibility are in direct line with the ones proposed to explain the small firm effect. Transaction costs are frequently mentioned (by, among others, Brigham and Smith, 1967, Neave and Wiginton, 1981, Sprenger et al., 1982 and Viscione, 1986), but marketability considerations also occur (Walker and Petty, 1978b). Interestingly, in the small business literature the control argument is usually described from the owner/manager's point of view. Owners of small firms may have an ingrained, strongly negative attitude towards going public, primarily because of aversion for accountability to outside stockholders and because of the wish to remain independent and flexible (Dailey et al., 1977). This may lead to constructions in which outside stockholders are granted very little control over the firm and this, in turn, may evoke the compensatory claims mentioned earlier.

### 2.4.4 *The small firm's capital markets*

The arguments in the previous subsection underscore the small firm's limited, or even absent, access to large scale capital markets like stock exchanges, and this conclusion is easily confirmed by the casual observation that no really small firms are listed. However, it would be premature to conclude from this that small firms consequently suffer from chronic undercapitalization, as is sometimes suggested (Tamari, 1980, Donckels, et al., 1987). Funds can also be raised on a smaller scale from regional, local or even personal sources, in "small scale" capital markets to which SME's (and consumers, for that matter) have ready access. SME's are almost entirely committed to these markets, without apparent detriment to their financing, since the literature does not reveal systematic scale effects in financial structure. Unfortunately, small scale capital markets are considerably less transparent than their large counterparts, and this problem is even aggravated by the confidential nature of the data involved in small business finance. Hence, any attempt to map these markets is likely to produce only partial results and will be compelled to leave large spaces blank, but some observations regarding the organization, infrastructure and market parties can nonetheless be made. Beforehand, however, it may be useful to illustrate how small financial operations in small business can be. Donckels et al. (1987, p. 19) report that almost three-fourth of the new enterprises in Flanders is started with a capital of less than BF 500,000, the value of a passenger car in the popular price range. New businesses in the USA are reported to have total capitalizations of $30,000 to $40,000 (The State of Small Business, 1986, p. 65). Although established small firms usually have higher capitalizations, it will be obvious that the amounts involved in financing a small business require a different approach than is offered by the stock exchange.

The first observation that can be made is that small scale capital markets are of the "negotiated" or "private placement" type, as opposed to the "auctioned" or "public issue" type stock exchanges. Since negotiated markets are by nature much more secluded than auctions, this aspects adds significantly to the intransparency of these markets. Note that most arguments for the small firm's limited access to capital markets are rooted in the way large auctions like stock exchanges are organized. Negotiated markets offer several possibilities to reduce transaction and information costs and to overcome marketability and control limitations. Potential investors can be made privy to information in a far less expensive manner than the preparation of a prospectus for a public issue. Additionally, some potential investors may be substantially more knowledgeable about the firm's prospects than investors at large. These differences in information, which may produce heterogeneous expectations as to the firm's prospects, can be strengthened by providing some investors with information that the firm would not be willing to make public because it would benefit the competitors. The disadvantage of limited marketability can be abated by close communication between the investor and the firm on a continuing basis and more or less frequent renegotiations of the terms of the investments. These terms can be tailored to the preferences of both parties and do not have to anticipate the desires of the market in general. The terms may also include all sorts of protective covenants to compensate the lack of control over the firm. When these possibilities are utilized, limited access to large scale capital markets does not have to affect the financing of the firm. Indeed, research by Dailey et al. (1977) shows that executives in smaller firms felt they had the advantage of easy access to modest amounts of capital funds. Moreover, most executives in family owned business, no matter how large their business, felt that they had a major advantage in acquiring funds because the relationship between family members and members of the lending institution was often a long and cordial one (Dailey et al., 1977, p. 32).

A second observation that can be made, is that the "financial infrastructure" of small scale capital markets is heavily debt oriented[13]. Only the main suppliers of debt, banks, have a widespread network of local branches. Besides a ready distribution channel, there is also a broad variety of legal instruments supporting the supply of debt. Interest payments are tax deductible. Contracts involving mortgages, fiduciary transfers of property, securities and liens, etc. are all worked out in detail by civil codes and are interpreted in extensive case law. Moreover, lending institutions have accumulated decades of experience, they have developed procedures for loan evaluation and credit monitoring and they maintain a highly trained staff. In short, the supply of debt is a profession, with its own traditions, codes, training facilities, collective agreements, etc. This extensive local, legal and professional infrastructure facilitates an efficient distribution of debt, even in small amounts, and a widespread use: if overdraft facilities on current accounts are included, it seems safe to say that all firms use debt in one way or another. For equity, on the other hand, the infrastructure is considerably less developed. The few organizations that supply equity on a professional basis certainly do not have a network of local branches. The legal forms of equity are, roughly speaking, either of the private partnership type, which mixes management and investment, or of the anonymous stockholder type. Neither is particularly suited to enable professional organizations to participate only financially in a large number of small firms. In The Netherlands, venture capital investments were rather cumbersome and received little attention until in 1981 an alteration of the law enabled banks and other organizations to start so-called private participation companies, which are partly government guaranteed. So the supply of equity is a comparatively new profession (for most banks at least) without tradition and substantial experience. Yet, all firms have equity. Apparently, a financial infrastructure is instrumental rather than essential to the small

-------

[13] *The argument here refers to the Dutch situation.*

firm's capital markets.

Reviewing the issue of the infrastructure, the impression arises that, in comparison with their large counterparts, the small firm's capital markets seem to be more pervaded by the possible conflict of interest between investors and the firm, and less focused on developing financial instruments that provide a well balanced mix of risk and return in various situations. The available instruments, often dating from several decades ago, typically provide security for lenders and control for equity holders; "mixed" instruments as convertible bonds or warrants have no equivalents in small scale markets. Despite the drastic changes in the economic and financial situation over the past decades, there have been virtually no financial innovations on small scale capital markets[14]. Nevertheless, many small firms, particularly start-ups, would benefit considerably if financial innovations would enable them to offer the suppliers of capital, especially equity, other terms of investment than the present combination of low taxable returns and high risk. For instance, it would be quite possible to develop a sort of "secured equity", with a posteriority claim on the firm's residual income and a priority claim on the proceeds of a distress sale of certain assets, e.g. buildings. This would relieve the firm of interest and repayment obligations, which can be a heavy financial burden for young, growing firms. It would also limit the potential loss of the suppliers of capital, while the risky nature of the investment would permit the amount of the investment to exceed the execution value of assets serving as security. Other financial instruments are also conceivable, such as the so-called "risk loan", a sort of convertible debt, which recently has been proposed in The Netherlands (see e.g. De Ridder, 1985). However, as long as the thinking of small business finance continues to be dominated by conflicts of interest instead of offering mature investment opportunities to mature investors, the chances of financial innovations are slim.

The third and last observation regards the market parties or the sources of financing available to small firms. The sophistication of the financial infrastructure is, of course, associated with the degree of organization and professionalism of these market parties, and not so much with the kind of capital supplied, as the above discussion might suggest.

-------

[14] *An exception is the above mentioned alteration of the law which enabled banks to enter the venture capital market by means of private participation companies. However, the effects of this, which are discussed later on in this study, are very limited.*

Table 2.4 Alternative sources of financing available to small business

| DEBT CAPITAL | start ups | SME's 78–80 | EQUITY CAPITAL | start ups | SME's 78–80 |
|---|---|---|---|---|---|
| | | INTERNAL | | | |
| Loans from owners/ shareholders | – | 6% | Capital: | | |
| | | | –individual savings | 13% | 2% |
| | | | –savings with spouse | 7% | |
| | | | Depreciation | – | 37% |
| | | | Retained earnings | 6% | 14% |
| | | | Other | 4% | 5% |
| | | INFORMAL EXTERNAL | | | |
| Loans from friends & relatives, etc. | 10% | – | Investments by friends & relatives, etc. | 4% | – |
| Trade credit | – | 11% | | | |
| | | FINANCIAL INTERMEDIARIES | | | |
| Commercial banks | 29% | 26% | Venture capital co's. | – | – |
| Savings & Loans | 2% | | Small business investment co's. | – | – |
| Governm. agencies | 14% | | Other | – | – |
| Finance & insurance co's, etc. | 11% | | | | |
| | | PUBLIC MARKETS | | | |
| Bond issues | – | – | Stock issues | – | – |

*Note Table* 2.4 :
*Compiled from The state of small business* 1984, 1986 *and* 1987. *Start-ups refer to the unweighted average of women and men owned businesses; trade credit is not included because of inadequate data (taken from Table* A2.10, 1986 *report). SME's* 78-80 *refers to the* 1978-1980 *average of funds invested in small corporations (Chart* 3.2, 1984 *report). The description of the sources comes largely from the* 1987 *report (Table* 2.1).

Table 2.4 presents an overview of alternative sources of financing available to small business in the USA, as well as some tentative data on the importance of these sources for business start-ups and investments in small corporations. These data indicate that debt is primarily supplied by professional organizations as banks, finance companies, savings and loan associations, etc. Equity, on the other hand, mainly comes from internal sources as depreciations and retained earnings or from informal sources as personal savings or friends and relatives[15]. Borrowing from the former sources

-------

[15] *Data of similar purport are provided by, among others, Walker and Petty* (1978*a*) *and Tamari* (1980).

clearly requires more infrastructure than tapping the latter sources. A few remarks to supplement the information in table 2.4 seem to be in place.

First of all, personal savings and friends and relatives appear to be an important source of capital for the starting small business. Indeed, virtually every book on small business finance mentions family, friends and relatives as a source of capital, and a collective term, Aunt Agatha, arose to name this community of anonymous investors (see e.g. Woodcock, 1986). Note that the use of this source of capital is facilitated by the very size of small business start-ups: investments amounting to the value of a passenger car are within the reach of many an Aunt Agatha[16].

The second remark is that the heavy use of internal and informal sources of finance induced some authors, of which Levin and Travis (1987) are probably the most outspoken, to question the value of the small firm's financial statements. Their argument is that the line between personal and business finance is blurred, so that elements of the financial statement may not be what they appear to be. Loans may in fact be easy to retrieve equity investments, assets may be inflated by beach cottages and mountain chalets and relatives may be employed and paid salaries well above the market rate, etc. Although these phenomena do indeed occur, their importance should not be overestimated for several reasons. First, they are not restricted to small business. In large business, complex holding structures occur in which the line between one company and the other is also blurred, with similar consequences for loans, assets, etc. Second, investors are usually not led astray by these tricks or constructions; e.g. banks ask security for their loans from either side of personal-business line. And last but not least, blurring the line between personal and business finance, or between debt and equity or between one company and the other, ultimately blurs the insight into the efficiency and profitability of the firm's operations. The firm's managers/owners stand to lose most from this blurred insight, so they have a clear incentive to keep this phenomenon well within limits.

The third and last remark referring to table 2.4 concerns venture capital, the importance of which is clearly limited in quantitative terms. The attention for venture capital grew considerably in the wake of the reappraisal of small business in general, and hi-tech firms in Silicon Valley were again the favourite examples. The latter may have been stimulated by success-stories like that of Apple Computer Inc., an earlier product of

-------

[16] *A small size also facilitates the use of colourful financial instruments, such as the innkeeper who pays dividends once a year in any percentage alcohol.*

venture capital investment, which was brought to the public issue market in 1980. Although success-stories of daring investments in operations that are started in a garage are appealing to the public at large, the importance of venture capital is best illustrated with some facts. In 1984, there were 509 venture capital firms in the USA, two-thirds of which managed a fund of less than $25 million. In 1983, $4 billion was committed to venture capital, and venture capital firms in turn invested $2,8 billion in about 1,000 companies. In an economy as large as that of the USA, these figures are practically negligible. In The Netherlands, the importance of venture capital is similarly small. In 1986 there were some 60 private participation companies, which participated in 211 companies with an average investment of *f*0.84 million per company (Koedijk and Kool, 1987). The total supply of venture capital in The Netherlands is estimated to amount to *f*1900 million (including the regional development companies as the N.O.M. and the B.O.M. and companies as the N.I.B. and the M.I.P.). See Schuit and Van Gelder (1987) for a detailed discussion of venture capital in The Netherlands.

### 2.4.5 *Conclusions*

So far, the harvest of size related characteristics of business finance is rather poor. Empirical studies show seemingly random size effects in financial structure, and the discussion of the small firm effect in stock pricing models produced many valuable suggestions, but no satisfactory explanation of the effect itself. A description of the small firm's capital markets did not reveal an abundance of scale effects either, although small firms were seen to raise their money in comparatively intransparent, small scale capital markets, with a heavily debt oriented infrastructure and without major financial innovations. The professional parties on these markets supply almost exclusively debt. A logical conclusion from these findings would be, that apparently there are not that many differences between small and large business finance. Somewhat conjecturally one might add, that although specific markets and financial instruments may be limited to a certain size range, the combined effects of these limitations cancel out, on balance. More specifically, and more conjecturally, one might say that as the size of transactions decreases, various marketability and cost considerations induce professional investors to gradually withdraw from the small firm's capital markets, but, enabled by the small amounts involved, informal investors successively take their place. There are, however, no clear indications that the latter investors are moved by an entirely different "invisible hand" than the former.

# PART II
# inventory of financial structure determinants

# 3 DETERMINANTS FROM THE THEORY OF FINANCE

## 3.1 Introduction and overview

The determinants of the firm's optimal capital structure have been a contentious issue in the theory of finance for three decades now and the debate, that was started by the publication of Modigliani and Miller's (1958, 1963) seminal papers, is still far from being concluded. As Modigliani and Miller's initial conclusions of capital structure irrelevance and, subsequently, of optimal capital structure at a 100% debt financing are clearly incompatible with observed capital structures, their findings initiated a considerable research effort to identify costs of debt financing that would offset its corporate tax advantage. Robichek and Myers (1965) and Baxter (1967) were among the first to suggest that the dead-weight losses associated with bankruptcy and reorganization might represent the major missing element. Since then, Kraus and Litzenberger (1973), Scott (1976), Kim (1978) and other authors, using a variety of approaches, have shown that the combination of leverage related costs and a corporate tax advantage of debt produces an optimal capital structure at less than a 100% debt financing, as the tax advantage is traded off against the likelihood of incurring the costs. However, both major elements in this "tax savings-bankruptcy costs" model became subject to criticism. This criticism is mainly of empirical nature and directed to the quantitative result of the model (i.e. the locus of optimal capital structure). Authors like Haugen and Senbet (1978) have argued that bankruptcy costs may be too small to have substantial influence on optimal leverage. Miller (1977) questioned the tax savings side of models of optimal capital structure, arguing that, under certain conditions, the corporate tax advantage of debt may be offset by tax disadvantages at the personal level, making leverage again irrelevant.

While the tax savings-bankruptcy costs model continues to engage researchers, other important developments in the theory have taken placc. The introduction of agency theory by Jensen and Meckling (1976) and of signalling theory by Ross (1977) into the field of finance enabled an explanation of capital structures apart from a bankruptcy costs - tax savings trade off. The importance of these extensions is not only that they facilitate an alternative way to reconcile observed capital structures with theoretical conclusions, but also, and perhaps more importantly, that they highlight different aspects of the complex financing decision. The general result of all these extensions of the Modigliani–Miller theory is, that the firm's optimal capital structure will contain a mixture of debt and equity if there is a positive net tax advantage to corporate debt financing and if there are significant leverage related costs, such as bankruptcy costs, agency costs of debt and loss of non–debt tax shields. The conditional nature of this conclusion springs from the recognition that the existence of an optimal capital structure is not only a theoretical issue, but also an empirical one whether or not the various leverage related costs and benefits are economically significant enough to have an appreciable impact on optimal leverage. This recognition gave rise to a number of empirical studies in which either the magnitude of leverage related costs and benefits is directly estimated (e.g. Warner, 1977) or in which observed capital structures are related to operating characteristics which are assumed to reflect these costs and benefits (e.g. Flath and Knoeber, 1980, Castanias, 1983, Bradley, Jarrell and Kim, 1984, Titman and Wessels, 1985). Although the hypothesis of a mixed capital structure is generally not rejected, the empirical evidence is yet too limited to fully appreciate the importance of the various advantages and disadvantages of debt financing. However, from the combined theoretical and empirical research efforts some broad categories of debt ratio determinants now gradually seem to emerge.

These determinants are generally described at a rather high level of abstraction, i.e. they are based on a large number of assumptions. For instance, capital markets are usually assumed to be costless, competitive and complete. Firms are assumed to separate investment from financing decisions, to issue only a very limited number of debt and equity categories and to maximize the welfare of their owners by maximizing the market value of the firm. Investors are assumed to act rationally, to be insatiable, to have homogeneous expectations regarding the future and to have equal and costless access to all relevant information. The combined result of these assumptions is a highly abstract financial world (with some bizarre consequences, e.g. slavery is allowed and no one can be fooled).

In spite of the high level of abstraction of the theoretical analyses, their empirical implications are usually formulated in terms of directly observable phenomena, primarily operating characteristics. Unfortunately, there is a wide choice of operating characteristics that may be used as proxy–variables for theoretical determinants. Hence, many variables have been put forth through time, ranging from straightforward variables like "depreciation charges" to less obvious concepts as "uniqueness". In addition to this variety, one operating characteristic can be linked with more than one theoretical determinant and conversely. The resulting jumble of theoretical determinants and empirical proxy–variables may at first sight seem to be somewhat confusing, but it can very well be summarized by distinguishing four categories of theoretical determinants as well as three broad categories of empirical proxy–variables. Theoretical determinants can be collocated in the categories taxes, bankruptcy costs, agency costs and signalling effect, while their proxies can be classified by the firm's assets, earnings and ownership structure. Table 3.1 shows the result of these two classifications for a number of operating characteristics that are expected to be related to differences in capital structure, and the associated references to the theoretical literature. The remaining part of this chapter is devoted to a brief description of the capital structure determinants in the present theory of optimal leverage and their empirical implications. Each determinant is briefly elucidated in the next sub–section and for each operating characteristic in table 3.1 the relation with theoretical determinants is expounded and a number of proxy–variables is presented in section 3.3.

Table 3.1 Overview of theoretical determinants of capital structure

| Theoretical determinant | proxy variable in terms of the firm's: assets | earnings | ownership |
|---|---|---|---|
| **Taxes** | –depreciation charges (DeAngelo & Masulis '80) | –non–debt tax shields (Bradley, Jarrell & Kim (1984) | |
| **Bankruptcy costs** | –liquidation value (Scott '76) | –size (Warner '77) | –complexity of ownership struct. |
| **Agency costs of**: | | | |
| –debt or equity | –collateral value (Jensen & Meckling 1976)<br>–growth (Myers 1977) | –volatility (Baxter 1967)<br>–profitability (Myers 1984) | –complexity<br>–co–ownership (Copeland & Weston 1983) |
| –customers etc. | –durables effect (Titman 1984) | | –limited liability/ complexity |
| **Signalling effect** | –private assets as collateral (Leland & Pyle 1977) | | –fraction of owners (or relatives') wealth held in the firm (Leland & Pyle 1977) |

## 3.2 Determinants

### 3.2.1 *Taxes and bankruptcy costs*

The influence of taxes on financial structure rests on differences in fiscal treatment between debt and equity: the fiscal regime is generally considered to favour debt. This simple statement reflects the ultimate result of the jungle of levies, subsidies, excises and the countless tax deductible items that make up the fiscal legislation in most western countries. However, the tax effects that are specified in theories of optimal capital structure are much simpler (perhaps because financial economists, like most people, need expert advice to complete their income tax forms). In essence, two major choices are to be made regarding the tax effects in theoretical exercises, viz. whether taxes are levied at both the corporate and the personal level or only at the corporate level, and what tax system is used.

As regards the latter, two basic tax systems occur in models of capital structure choice, the wealth tax system and the income tax system. According to the wealth tax system, all payments to debtholders, including principal repayments, are deductible from the firm's taxable income, but depreciation charges and other non–debt tax shields are not. The income tax system allows the deduction of interest payments, depreciation charges and other non–debt tax shields, but principal repayments are not deductible. Obviously, the income tax system more closely resembles the tax systems in force in most western countries than does the wealth tax system. On the other hand, a wealth tax system is much easier to implement in models of capital structure choice (see also section 4.2). Both systems are frequently encountered in the literature.

As regards the other major choice, the introduction of personal taxes in addition to corporate ones is not likely to be neutral with respect to the firm's financing decisions, as Miller's (1977) stirring article clearly demonstrates. Miller specifies personal taxes so, that they offset the favourable tax treatment of corporate debt. Hence, when both are integrated within a general equilibrium context, the tax incentive for corporate leverage will disappear altogether for any individual firm, even though there will still exist an optimal debt–equity ratio for the corporate sector as a whole. Although irrelevance is an extreme implication of the Miller argument, it counterbalances the extreme implications of the original Modigliani–Miller theory and it illustrates the possible impact of personal taxes.

Personal taxes can also have other effects. In many countries, the tax rates applying to capital gains are less than the rates applying to dividend and interest income, and capital gains taxes are levied only when the gains are realized by selling the asset. This creates a preference for capital gains over dividends from stock ownership on the part of individual investors which can, in turn, produce a systematic effect on firm financial decisions. However, this effect may be cancelled out by other aspects of the tax environment. Some investors are not taxed, e.g. very large investors like pension funds or very small investors who benefit from the fact that small amounts of dividend are usually exempt from taxation. Other investors, e.g. retired people, may depend on dividend income for their livelihood. Given the wide variety among investors' tax situations, it is impossible to say a priori which policy should be pursued. It is likely, however, that each policy will attract a certain "clientele" of investors, whose preferences correspond with the characteristics of that particular policy. This clientele effect, as it is called, does not imply that one policy is better in terms of firm valuation than any other, but it does imply that firms should establish and maintain a specific policy, making few changes over time. Firms that fail to maintain a specific policy may cause current shareholders to incur unwanted transaction costs; such firms are unreliable and, therefore, relatively unattractive to all investors, regardless their preferences. (See Copeland and Weston, 1983, for a discussion of empirical tests of the clientele effect).

If all the above effects would be accounted for, the tax effects in theories of optimal capital structure would become too complex to handle (and financial economists would need expert advice to complete their own models). However, the net effect of all fiscal rules is usually assumed to create a positive tax incentive for corporate debt financing, albeit that this effect is smaller than in the original MM theory and it may not apply equally to all investors.

<u>Bankruptcy costs</u> refer to the added costs (such as legal fees, reduced sales, increased production costs, etc.) that arise because the firm cannot meet its obligations to creditors without changing its operating or external financing activities (Haley and Schall, 1979). In this description, bankruptcy is more general than the corresponding legal concept and some of the costs referred to above may arise without the firm actually being declared bankrupt. A mild form of "financial distress", as Haley and Schall call it, exists if the firm cannot make its debt payments without curtailing

investment or resorting to external financing. Bankruptcy costs[1] influence financial structure because they reduce the value of the firm by the fact that additional costs arise. An increased probability of bankruptcy, resulting from e.g. increased interest payment obligations, provides an increase in the expected bankruptcy costs, which discourages the use of debt. This fact can be used to explain the existence of an optimal capital structure.

From a theoretical point of view, bankruptcy costs are almost part and parcel of models of optimal leverage. Empirically, however, their importance is somewhat controversial. The first evidence, collected by Warner (1977), indicates that direct bankruptcy costs, such as lawyers' and accountants' fees, are trivial, averaging about 1% of the market value of the firm prior to bankruptcy. Furthermore, these costs seem to decrease in percentage terms with increasing firm size. This would make bankruptcy costs a more important capital structure determinant for small firms than for large ones. On the other hand, Kim (1978) quotes evidence in which total administrative expenses in a business bankruptcy are estimated to approximate 20% of the estate. He also mentions that some costs decline with firm size, stressing the importance for smaller firms.

However, direct costs may be the most obvious part, they are probably also the least significant component of the costs associated with bankruptcy. Kim (1978) distinguishes the additional components "indirect costs" and "loss of tax credits"[2]. Substantial indirect costs can arise from the imperfection of secondary markets for physical assets[3]. To illustrate, Kim mentions that in a distress sale, assets may bring as little as 30% of their wholesale price (Kim, 1978, p.47). Other indirect costs arise, when customers turn to other firms, because they question the reliability of the firm as a supplier due to problems the firm has in meeting its current debt obligations. Similar costs arise when the firm's management and other employees start looking for other jobs or want to renegotiate their contracts. Creditors may also want to

-------

[1] *Haley and Schall's terminology ("costs of financial distress") is probably more appropriate but "bankruptcy costs" became the current term for these costs*

[2] *See Jonkhart* (1980, *p.*29) *for an enumeration of various possible elements of bankruptcy costs.*

[3] *Note that these costs are, strictly speaking, associated with liquidation rather than bankruptcy, and the one does not necessarily imply the other. However, if assets are dismantled and sold to meet obligations to debtholders the loss in value is a cost of financial distress, even without bankruptcy actually taking place.*

renegotiate, thus forcing the firm's management to spend considerable time negotiating with employees and creditors or their lawyers. This, of course, reduces the time available for normal responsibilities. Furthermore, the firm may experience difficulties in financing its investments so that profitable opportunities must be forgone. All these effects contribute to a reduction in firm value if the possibility of "financial distress" exists, even if bankruptcy in the legal sense is virtually impossible. Hence, increasing the expected value of these costs by choosing high debt levels which increase the probability of bankruptcy will, at a certain point, cancel out the tax advantage and thus produce an optimal capital structure.

The discussion of indirect bankruptcy costs contributes significantly to our insight into the consequences of financial distress. However, the empirical problem is not alleviated by it. There are no reliable estimates of the magnitude of these indirect costs. So, taken as a whole, the empirical evidence on the magnitude of all kinds of bankruptcy costs is mixed and too limited to fully appreciate their importance, but it is large enough to grant them a place in the theory of capital structure.

### 3.2.2 *Agency costs and signalling effects*

The elements of the tax savings–bankruptcy costs model, which are discussed in the previous paragraph, are not only vulnerable to criticism from an empirical point of view. As Copeland and Weston (1979) point out, firms in the USA used debt in their capital structure before income taxes existed and the cross-sectional regularities in financial leverage, that existed before corporate income taxes were levied, were not upset by the introduction of these taxes. This suggests that other costs and benefits also play a role in the determination of capital structure.

Perhaps the most powerful concept of leverage related costs was introduced by Jensen and Meckling (1976), who describe the firm as a nexus of relationships that can be characterized as principal–agent relationships. An investor who enters into a financial contract with a firm (or an entrepreneur) can be thought of as employing the firm to manage his assets. Thus, the firm's security holders (debtholders as well as stockholders) can be seen as principals and the firm's management as the agent. As Jensen and Meckling point out, this principal–agent relation is costly, because there is a possibility that the agent will not always conduct business in a way that is consistent with the best interest of the principals. Hence, the latter may insist on various types of protective covenants and monitoring devices in order to protect themselves. Different investors are likely to rely on different kinds of monitoring. Lenders will monitor the firm's operations to prevent the borrower from taking excessive risk, because risk-taking tends to benefit the borrower more than the lender. Even when debt is secured by assets such as buildings and equipment, collateral can be dissipated through inadequate maintenance and excessive payments of dividends and salaries to owners. For equity holders, monitoring activities will concentrate on the entrepreneurs' actions to prevent them from taking benefits at the expense of the shareholders.

Monitoring is generally facilitated by using covenants that restrict the entrepreneur's activity, thereby increasing financing costs. Contract provisions may prohibit a borrower from incurring additional debt, from paying dividends and from removing assets from the business site. Venture capital firms frequently reserve the right to install new managers in the firms in which they invest. However, the costs of writing and enforcing these covenants and of employing other monitoring devices may well be substantial, and these costs are likely to increase with the amounts of money supplied by the security holders. Therefore, security holders will not seek perfect protection, but these agency costs, as Jensen and Meckling call them, will be traded of against the likelihood of a loss resulting from management actions. If agency costs increase with the proportion of financing supplied by any type of security holder, conformably to Jensen and Meckling's suggestion, it is possible to argue that a capital structure exists in which total agency costs are minimal. In this way, an optimal capital structure may exist even in the absence of taxes and bankruptcy costs.

Monitoring costs are likely to be comparatively high for small and young firms, because the lack of formal financial control and the firms' flexibility to change its assets make it difficult to gain insight into their financial performance. However, once the firm is established, the costs of repeated debt financing from financial intermediaries can decline significantly because information is already available and a relationship with the lending institution has been established (cf. section 2.4.4).

The power of the concept of agency costs is that it need not be restricted to the firm's security holders, but that it is applicable to a broad variety of situations: the firm can be pictured as a web of relationships between all participants and stakeholders (see Copeland and Weston, 1983, ch. 13 for an overview). For instance, agency costs can play a role in the contracts between the firm and its customers if the firm sells durables that require maintenance in future periods (Titman, 1984). Similarly, the relationship between the firm and its employees may be described in terms of a principal–agent relationship. Particularly the organizational and "pecuniary" aspects of the agency relationship between owners and managers received considerable attention in the literature (Williamson, 1975, is a prominent example of research in this field). An organizational implication is, for instance, that the separation of "decision management" and "decision control", which is found in most large corporations, can be an efficient way to reduce agency costs (Fama and Jensen, 1983). Another implication is that, given the difficulty of monitoring the activities of managers, it is important to motivate managers to act in the best interest of the owners. Incentive schemes, such as stock option plans, can create this motivation. However, they solve the problem only partially since they expose the managers to the unsystematic risk of the firm by reducing their opportunities to diversify their personal investments. This could induce managers to make decisions that are less risky than would be optimal from the owners' point of view. These aspects of the agency theory are less relevant in small business because the vast majority of small firms is managed by its owners. However, the nondiversification argument (reached in a somewhat roundabout way) is, of course, also valid for owner/managers. In fact, the personal investments as well as the "human capital" of owner/managers can be tied entirely to the fortunes of their firm. An implication of this nondiversification argument is that closely held firms (not necessarily small) may use less debt than would be optimal from a theoretical point of view. Another implication is that characteristics of the owner/manager, such as his attitude towards risk on his personal wealth outside the firm, may be reflected in the firm's financial structure.

In addition to the foregoing it can be noted that some, if not all, of these relationships have information value, as is suggested by Ross' (1977) signalling hypothesis. The financial obligations a firm enters into may serve as signals to the public about the future performance of the firm. These signals cannot be mimicked by weaker firms because they lack the cash flow to meet these obligations. Dividend policy is another obvious candidate for a signalling device. Still other signals may spring from the owner/manager's personal financial involvement in the firm, as is suggested by, among others, Leland and Pyle (1977). They argue (p.371) that borrowers typically know their collateral, industriousness and moral rectitude better than do lenders; owner/managers possess "inside" information about the projects for which they seek financing. On this basis, Leland and Pyle develop a valuation schedule in which the value of a project depends on the fraction of equity retained by the owner/manager. This fraction, like the obligations and dividends mentioned above, serves as a signal to the firm's potential security holders indicating the quality of the firm's investment projects. Alternative signals may be the fraction of the owner's wealth held as equity in the firm[4], or the participation of closely related relatives in the firm.

Signalling activities arise in situations in which information is unevenly distributed over investors and management (i.e. when "information asymmetry" exists) and when moral hazard hampers the direct transfer of information between market participants. Generally, the firm's managers are most knowledgeable of its prospects, but they cannot be expected to be entirely straightforward about these prospects for two reasons. First, there may be substantial rewards for exaggerating the firm's positive qualities and obscuring the negative ones, when ascertaining the true qualities by outside parties is costly or impossible. Second, managers can be reluctant to disclose the full merit of the firm since this might allow competitors to appropriate key features of the firm's projects.

Information asymmetry is particularly pronounced for, again, small and young firms in a dynamic industry. These firms can incur high costs to generate information for use by outside investors, because much of their expertise and knowledge is intangible and hard to quantify. So, like bankruptcy and monitoring costs, information costs are also likely to be a more important financial structure determinant for small businesses than for large ones (cf. the small firm effect in section 2.4.3)

## 3.3 Empirical implications

### 3.3.1 *Implications pertaining to the firm's assets*

The past investment decisions of the firm, materialized in the assets it owns, play a role in many theoretical contributions, and the firm's capital structure may indeed be related to several characteristics of its assets. The first and perhaps most obvious characteristic that may affect the capital structure choice is the liquidation value of the assets, as is formally shown by Scott (1976). Bankruptcy costs can be thought of as the difference between the firm's operating value and its liquidation value, so a high liquidation value makes debt financing more attractive by reducing bankruptcy costs. The

-------

[4]*It should be noted that the desire to diversify will increase with the wealth of business owners, because the cost of nondiversification, i.e. the allocation of a large proportion of wealth to a single investment, can be high. This stresses the value of the signal.*

collateral value of the assets is closely related to the liquidation value (in fact, both terms often are used interchangeably). However, collateralizing a loan requires a well-defined covenant in the debt contract and ties the funds of a specific lender to a specific investment. In addition to reducing bankruptcy costs, this also diminishes or eliminates the risk that the funds are used for another, more risky, investment than originally specified. This, in turn, reduces the need to incur monitoring costs in order to check management actions, which makes debt financing less costly.

An indirect effect of the asset structure may be caused by differences in fiscal treatment (e.g. regarding depreciations) between categories of assets. As the assets offer more possibilities to avoid tax-levy, the tax advantage of debt financing may decline (see also section 3.3.2).

A special asset-category distinguished in the literature is constituted by the firm's growth opportunities, which can be considered assets that do not generate current taxable income, that cannot be collateralized and depreciated and that offer the firm's management a high degree of flexibility in selecting future investments. With the exception of the depreciation effect, these arguments suggest that debt levels are negatively correlated with growth. However, as Myers (1977) points out, the agency problems caused by the lack of collateral and high flexibility of growth can be abated by resorting to short term debt, which offers the possibility to frequently renegotiate the terms of the debt contract. Hence, growth may have a positive influence on short term debt, although the influence on total debt is expected to be negative. It should be noted, however, that growth can also be considered a signal of the quality of the firm's investments (as is commonly done in practice). In that case, the expected influence on debt is positive.

Another special case concerning assets occurs when business loans are collateralized with private assets. Besides reducing bankruptcy and agency costs, as business collateral does, this has the additional effect of serving as a signal about the quality of the firm's investments (based on the entrepreneur's "inside information"). Therefore, we would expect private collateral to have a greater impact on leverage than business collateral. Although the influence of private collateral is scarcely mentioned in the literature, it is brought forward here because it is a common phenomenon in small business. The reason for this is, that it is easier to place a value on a personal property, such as a residence, than on the human capital which usually constitutes a large part of the small firm's income generating assets.

Finally, as Titman (1984) argues, debt ratios may depend on the type of products a firm carries. If a firm sells durables that require maintenance in future periods, the customers buy a future stream of services as well as the product itself. These services are lost upon bankruptcy and, consequently, customers may reclaim a part of the purchasing price, thereby diluting the debtholders' claims. Additionally, firms that increase the probability of bankruptcy by carrying more debt will be confronted with a lower demand for their products. Thus, firms involved in durables are expected to carry less debt.

The empirical proxy-variables for the tax effect of the assets usually are ratios comprising depreciation charges, e.g. depreciation-to-operating income (Titman, 1983). The liquidation and collateral value effect generally are combined and measured by relating some asset categories (e.g. tangible assets, inventory and gross plant) (Titman and Wessels, 1985), fixed assets (Marsh, 1982) to total assets. Straightforward measures of growth are the change in total assets or sales, but other variables (e.g. research and development-to-sales) (Titman and Wessels, 1985) are also used. The durable effect is commonly captured by some industry classification, while private collateral can only be measured by the incidence or the amount of such collaterals.

### 3.3.2 *Implications pertaining to the firm's earnings*

An important characteristic of the firm's earnings relating to capital structure choice is the degree to which they are taxable. Tax deductible items other than the interest payments reduce the firm's taxable earnings and, hence, increase the probability that the tax advantage of debt is lost when the firm experiences low earnings. Therefore, firms with a comparatively large amount of non-debt tax shields will carry comparatively little debt, as is shown by DeAngelo and Masulis (1980) and Bradley, Jarrell and Kim (1984).

Size may be measured in terms of earnings, assets, employees, etc. which are all likely to be highly correlated. Size influences leverage if bankruptcy costs are a decreasing function of firm size, as the evidence presented by Warner (1977) and Kim (1978) indicates. Other influences of size occur when the transaction costs of debt and equity respectively are different functions of size and when the probability of bankruptcy is size dependent. However, the theoretical as well as the empirical arguments for size effects on capital structure are yet comparatively weak.

The hypothesis of a relation between the volatility of the firm's earnings and its capital structure has intuitive appeal and this hypothesis was among the first to be suggested. However, theoretical analyses can demonstrate an ambiguous influence of earnings volatility on optimal leverage, as will be shown in the next chapter[5], while empirical analyses show similarly mixed results (see Van den Bergh et al., 1985, for a discussion). In spite of this, volatility measures are frequently included in empirical studies as well as in summaries of the theory (see e.g. Myers, 1984, p. 581).

The importance of internally generated funds is stressed by Myers (1984) and Myers and Majluf (1984), following the observation that firms seem to prefer raising capital by retaining earnings in the first place, by borrowing in the second place and by issuing new equity in the last place. By retaining earnings a firm can economize the flotation costs of issuing new stock or bonds, but this only affects capital structure if these transaction costs are large compared with the disadvantages of deviating from the optimal capital structure. Another explanation of this "pecking order" in financing alternatives, as Myers calls it, results from an asymmetrical distribution of information between potential outside stockholders and the firm's management if the latter tries to maximize the "true" value of the existing shares. In this situation the firm will only issue new stock if it adds value to the existing shares, i.e. if the issue brings in more than it is worth. On the other hand, potential stockholders know this and will rationally adjust the price they are willing to pay. Although the functioning of the capital market under these circumstances remains somewhat unclear, it may result in a situation in which firms face the dilemma of either passing by projects with a positive net present value or issuing stock at a price they think is too low. This situation may be avoided if a firm can retain enough internally generated cash to cover its positive NPV opportunities or if it can maintain financial slack in the form of "reserve borrowing power". In this view, observed debt ratios will reflect the cumulative requirement for external financing over an extended period, and will be negatively related to profitability.

The characteristics of the firm's earnings pertaining to the tax effect usually are approximated by depreciations and investment tax credits related to operating income (Bradley, Jarrell and Kim, 1984) or to total assets (Titman and Wessels, 1985), but direct estimates of non-debt tax shields also occur. Volatility is commonly assessed by some statistical measure of dispersion of the earnings, such as the standard deviation of the

-------

[5]*See also Castanias* (1983) *and Van der Wijst* (1986)

standardized growth in sales (Ferri and Jones, 1979), standard deviation of changes in EBIT (Marsh, 1982), etc. The proxy-variables for profitability include ROI (Marsh, 1982), operating income-to-sales or to total assets (Titman and Wessels, 1985), and, as a measure of cumulative profitability, accumulated retained earnings-to-total equity.

### 3.3.3 *Implications pertaining to the firm's ownership structure*

The firm's ownership characteristics are not frequently encountered in the literature as possible determinants of capital structure. However, most, if not all, theoretical determinants are somehow related to the ownership structure and although some aspects of it are hardly accessible for outside observations, other aspects may well be included in empirical analyses.

The complexity of the ownership structure may influence bankruptcy costs as it is more costly to disentangle complex conglomerates than it is to liquidate a single firm. In this respect, complexity is not an exclusive characteristic of large firms; today, even small retailers frequently have separate corporate bodies for their real-estate and for their trading activities. The position of lenders, suppliers and employees may seriously be affected by these legal constructions, as e.g. seemingly substantial collateral may appear to be owned by a different corporation. Of course, the agency costs to reveal these constructions and to bypass them in (debt-)contracts may be considerable.

Another aspect of ownership structure that may influence leverage is co-ownership, which, according to Copeland and Weston (1983), implies agency problems. Co-owners will have to incur monitoring costs in one form or another in order to ensure that the firm is run in their best interest. Co-ownership may also cause conflicts of interest between the partners regarding e.g. paying dividends or retaining earnings. To avoid these problems, co-owned firms may rely more heavily on debt.

The third aspect of ownership structure that can be related to financial behaviour is the distribution of ownership. In closely held firms, owners can lack diversification opportunities, which may induce them to reduce the risk of bankruptcy by using less debt compared with firms that are owned by many (anonymous) investors. Moreover, the preferences (or peculiarities) of the few owners of a closely held firm may be reflected in it's financial structure (e.g. personal wealth outside the firm or attitude towards risk).

As was mentioned before, the ownership structure may also signal information to the capital market. Owners usually have inside information on their investments and the fraction of their personal wealth they are willing to commit to the investment for which they seek financing conveys information as to the quality of these investments. Hence, large personal commitments may be associated with high debt levels.

The complexity of the ownership structure is hard to measure in detail, but it may be approximated by the presence of more than one corporate body for one firm. Co-ownership typically influences the legal form (e.g. limited partnership), but it may also be reflected in the pay-roll (more than one director's salary). The distribution of ownership can be measured by the number of owners, but the nondiversification effect essentially refers to the difference between closely held and listed firms. In the latter case, the nondiversification effect for owners can be replaced by the same effect for managers, so that the empirical implications may be limited. The personal preferences of the owners of closely held firms are seldom available, but the possibility of usable proxy variables cannot be excluded beforehand. For instance, the attitude towards risk may be related to age, as is sometimes argued. The fraction of the owners' wealth held in the firm is again hard to measure, but usable proxy-variables are the number or the amount of collaterals, especially private ones, supplied by the owners.

## 4.1 Introduction

In this chapter, theoretical aspects of the debt financing decision are studied by constructing some models of capital structure choice. In these models, major recent developments in the theory of optimal capital structure are systematically restated in mathematical form, based on a uniform framework of definitions and derivations. The purpose of this procedure is twofold. First, by incorporating the various contributions into a uniform framework, distracting differences in methodology between them are eliminated, which facilitates a clear overview and a proper assessment of the developments. The second purpose is to provide an explicit valuation of corporate claims in every model that is investigated. As the interest of some authors centres on the existence of an optimal capital structure, this explicit valuation is not always adequately dealt with. Particularly the question of corporate debt capacity, which is defined as the maximum amount of debt the capital market is willing to provide, is neglected in a part of the literature on the existence of an optimal capital structure. This is a serious omission, because if debt capacity is reached before optimal capital structure, the theoretical optimum would not be attainable; the best attainable capital structure would coincide with debt capacity. Thus, the question of an optimal capital structure would be irrelevant, or in Myers and Pogue's (1974) terms, the lenders would chicken out before the shareholders. In this study, valuation formulae for debt, equity and their sum, i.e. firm value, are derived and the firm value maximizing degree of leverage (optimal capital structure) is calculated, as well as the degree of leverage that maximizes the value of debt (i.e. corporate debt capacity). Further, a comparative statics analysis is performed for the models that allow such an analysis. The comparative statics provide a basis for the formulation of empirically testable hypotheses and are therefore particularly valuable for the empirical analyses later on in this study.

Of course, summaries of theoretical developments have been published before; the studies by Chen and Kim (1979) and Jonkhart (1980) are good examples. Compared with the Chen and Kim summary, this study covers a wider area, including multi-period models and informational differences and it provides a rigorous analysis of each model presented. In the Jonkhart study, single-period models are presented in a different framework than the one used here. Both the Jonkhart study and the Chen and Kim study are aimed at presenting theoretical developments in a uniform framework. This aim is also pursued here, but an additional aim is to investigate the robustness of the various theoretical contributions, i.e. to ascertain how sensitive (published) models are for small changes in their specification. The latter area is still far from being completely mapped and establishing the robustness of theoretical models could in itself be a contribution to the theory.

Because the purpose of the theoretical exercises is to inventory financial structure determinants, we are primarily concerned with the qualitative results of the models. By qualitative results we mean the existence of optimal capital structure and corporate debt capacity and the direction in which these optima shift under the influence of changes in their determinants. The quantitative result refers to the locus of optimal capital structure respectively corporate debt capacity and the degree to which these optima change as a result of changes in the value of the determinants.

The framework adopted for this restatement is the extended Modigliani–Miller theory, modelled within the so–called mean–variance approach under a set of fairly restrictive assumptions. This particular framework is chosen because it is comparatively simple and because, over the years, it has been used to model a considerable part of the contributions to the theory. These characteristics fit in well with the purpose of the theoretical exercises, which is to inventory financial structure determinants in the literature rather than to advance the theory itself. However, the model construction efforts may contribute to the theory by throwing some light on the robustness of the various theoretical models. Restating these models within a single framework should, in principle, produce a tree–structured family of interrelated models, obtained by subsequently relaxing assumptions or introducing new phenomena. However, in view of the general tendency to publish only well–behaved models, this may not be the case, especially not as regards the "intermediate" models, i.e. non–published variants of well–known models.

The discussion in this chapter is developed in a step–by–step way, starting with the MM–theory. In each subsequent step one or more assumptions is (are) relaxed, highlighting a different aspect of the leverage decision, and the effects on the valuation of corporate claims are analysed. The analysis is, of course, restricted in several respects. Although a single–period setting as well as a multi–period setting are used, several other assumptions, such as discrete time and risk indifference, are maintained throughout the analyses. This benefits the comparability of most contributions to the theory, although it may exclude some recent developments. Further, derivations and proofs and the elaboration of some models which produce indeterminate results are relegated to a separate publication, in order not to distract attention from the valuation formulae and conclusions by overloading this study with mathematics. For extensive mathematical treatments of all models presented here the interested reader is referred to Van der Wijst (1986).

## 4.2 Assumptions and definitions

The assumptions made to facilitate the construction of the models attempt to capture, as simply as possible, the essence of the economic phenomena they seek to describe. Hence, no detailed technical realism is aimed at. To illustrate, consider the tax environment. In the previous chapter, two basic tax systems were introduced, the wealth tax system and the income tax system. Recall that the wealth tax system allows the deduction of all payments to the debtholders, including principal repayments, from the firm's taxable income, but depreciation charges and other non–debt tax shields are not deductible. The latter two are deductible according to the income tax system, together with the interest payments, but principal repayments are not. As is already mentioned, the income tax system is clearly more realistic than the wealth tax system. However, the latter is easier to implement in theoretical models, because it is not necessary to distinguish between interest payments and principal repayments, and the additional term(s) for depreciation charges and other non–debt tax shields can be omitted. So, if realism would be aimed at, the tax environment would involve corporate income taxes as well as (progressive) personal income taxes and numerous tax deductible items. However, the qualitative results of the models do not depend on the complexity of the tax system; the essence of taxes in theoretical models is that they induce an advantage of debt financing and this effect can be established, in qualitative terms, by a very simple tax system as well as by a more complex one. The quantitative results are, of course, affected by the tax system as well as by the tax rates, but in this study we are interested in influences on, and not the locus of, the value of corporate claims. Furthermore, it should be noted that the results of the analyses can be interpreted in broader terms than the assumptions they are based on. For instance, a very simple tax structure can be thought of as the net taxes that, over a certain range, result from a more complex tax system, involving both corporate and personal taxes. Also, single-period models can be interpreted in multi-period terms. For instance, the variables in single-period models can be seen as vectors over time. Alternatively, the end of period variables in a single-period model can be thought of as including the present values of these variables in future periods. So the assumptions are not as restrictive as they may seem at first glance.

Capital markets are assumed to be costless and competitive. In costless capital markets there are no transaction costs and financial assets are costlessly and infinitely divisible. In competitive markets, no buyer or seller is large enough to individually influence prices. Corporate profits are taxed at a fixed rate and personal taxes are neutral, i.e. there are no personal taxes or the effective tax rates on interest dividends and capital gains (realized or unrealized) are equal. The tax system used is the wealth tax system described earlier. The assumptions regarding taxes result in a tax advantage of debt financing, expressed in a simple way that facilitates a convenient notation. Further, all market participants are assumed to be insatiable and to act rationally.

The firm is considered at a point in time at which the investment decisions have already been made, whereas the financing decision has not yet been made. Hence, the firm's set of income generating assets (comprising both the physical assets in place and the future investment opportunities, if any) is assumed to be fixed. Firms only issue debt and equity; debtholders are protected against unanticipated transfers of their wealth to stockholders by protective covenants in the bond indenture. Debt claims are only subject to the risk of default; other risks, e.g. price level risk, are not explicitly dealt with in this study. The firm's debt capacity is defined as the maximum amount of debt the capital market is willing to provide. This means that, when debt capacity is reached, the firm cannot borrow more by promising larger payments to debtholders, i.e. debt capacity is reached at the degree of leverage that maximizes the value of debt[1]. The firm's optimal capital structure is defined as the degree of leverage that maximizes the wealth of the (existing) shareholders, which is, in most instances, equivalent to maximizing the value of the firm[2]. The firm's management is assumed to act in the best interest of the existing shareholders.

-------

[1] *See Kim* (1978) *and Jonkhart* (1980) *for a more detailed discussion of the concept of corporate debt capacity.*

[2] *In some instances, maximizing firm value is not equivalent to maximizing the existing owners' wealth* (*see e.g. Haley and Schall*, 1979, *p.* 477–82). *These instances are discussed in section* 4.4.1.

Investors are assumed to be indifferent to risk. Although several approaches to incorporate risk aversion into models of capital structure are available (state preference approach, value additivity principle), the –less general– capital asset pricing model (CAPM) would, for this study, be the most amenable model. Kim (1978) provides an analysis of the leverage decision based on the assumption that risky assets are priced according to the CAPM. However, in the position taken here, the firm's set of income generating assets and, hence, the distribution of the firm's cash flows, is assumed to be fixed. In the CAPM, risk is measured as the covariance between the firm's cash flows and the returns from the market portfolio, so the firm's risk characteristics are also fixed. In this position[3], the qualitative results of the analyses are unaffected by the assumption of risk neutrality, as can be shown by substituting a zero risk premium into the results of Kim's (1978) analysis. The incorporation of risk aversion would, however, add considerably to the complexity of the valuation formulae and seriously hamper the derivations. Therefore, the assumption of risk neutrality is maintained throughout the analyses. An additional effect of this procedure is, that it can provide an insight into the robustness of the various models under a set of restrictive assumptions which, in turn, may throw some light on the prospects of extending the models with more realistic assumptions.

## 4.3 Single period models

### 4.3.1 *Introduction: the MM – position*

In this section, the financing decision is analysed in a one period–two moment context. In this setting, decisions are made at the beginning of the period and the results become known at the end. Because there is only one period, payments to debtholders comprise both interest and redemption of the principal sum. Under the tax system introduced in the previous section both are tax deductible. In addition to the assumptions discussed in section 4.2, which are maintained throughout the analyses, the following assumptions are made in different combinations in different models. In each subsequent model, one or more of these assumptions is (are) relaxed, which highlights a different aspect of the leverage decision, and the effects on the valuation

-------

[3]*Note that the position taken here involves the the valuation of one firm with fixed risk characteristics and not e.g. the comparison of two firms with different risk characteristics. In the latter case, the assumption of risk neutrality may influence the qualitative result.*

of corporate claims are analysed. The first of these assumptions is that investors share the same subjective expectations regarding future events (i.e. they have homogeneous expectations). Further, all market participants are assumed to have equal and costless access to all relevant information. Bankruptcy is assumed to be costless. Finally, it is assumed that the firm's owners have unlimited liability and that their personal wealth is sufficient to meet the contractual obligations to the debtholders if these cannot be met out of the cash flow. The result of the latter assumption is that debt is riskless and that the stockholders always receive the, positive or negative, cash flow after interest and taxes. In combination with the foregoing, this is equivalent to the MM tax case. The valuation formulae for this case can be derived as follows. If we assume that the fiscal authorities pay a tax rebate if the firm suffers a loss, the after tax end of period value of equity, $Y_e$, is equal to[4]:

$$Y_e = (1-\tau)(\tilde{x}-R) ,$$

in which $\tau$ is the corporate tax rate, $\tilde{x}$ is the random variable representing the firm's cash flow before interest and taxes and $R$ represents the contractual payments to the debtholders. For risk neutral investors, the equilibrium value of equity, $V_e$, is the present value, discounted at the risk free interest rate, of the expectation of $Y_e$:

$$V_e = \frac{E(Y_e)}{(1+r)} = \frac{(1-\tau)\int_{-\infty}^{\infty}(\tilde{x}-R)f(\tilde{x})d\tilde{x}}{(1+r)} ,$$

in which $f(\tilde{x})$ is the probability density function of $\tilde{x}$. This expression can be written as :

$$(4.1) \qquad V_e = \frac{(1-\tau)(\bar{x}-R)}{(1+r)} .$$

The end of period value of debt, $Y_d$, can be derived in a similar way and equals the payments to the debtholders:

$$Y_d = R.$$

For risk neutral investors, the equilibrium value of debt , $V_d$, is the present value of the expectation of $Y_d$:

---

[4] *An overview of the symbols used throughout the analyses is given in table 4.1A in the appendix to this chapter.*

$$V_d = \frac{E(Y_d)}{(1+r)}, \text{ which is :}$$

$$(4.2) \quad V_d = \frac{R}{(1+r)}.$$

The value of the firm is found by adding up the values of equity and debt ($V = V_e + V_d$):

$$(4.3) \quad V = \frac{(1-\tau)\bar{x} + \tau R}{(1+r)}.$$

The first term in the numerator of (4.3) represents the value of the unlevered firm ($V_u$), while the second term represents the tax advantage of debt, i.e. the well known MM result $V = V_u + \tau V_d$. Differentiating (4.2) and (4.3) with respect to $R$ produces the well known MM results:

$$(4.4) \quad \frac{\partial V}{\partial R} = \frac{\tau}{(1+r)} > 0, \qquad \text{and} \quad \frac{\partial V_d}{\partial R} = \frac{1}{(1+r)} > 0,$$

which implies that the level of payments to debtholders that maximizes the value of the firm (optimal capital structure) is reached at a 100% debt financing. Similarly, the level of payments to debtholders that maximizes the value of debt is also reached at a 100% debt financing. Although this case is too well known to need further explanation, it is presented here as a natural starting point of the analyses.

### 4.3.2 *The effect of risk of default and limited liability*

The influence of the risk of default on the valuation of corporate claims is closely connected with the influence of limited liability; they are, so to speak, opposite sides of the same coin. In the previous subsection, debt was not subject to the risk of default because stockholders were assumed to have unlimited liability and sufficient personal wealth to meet the contractual obligations to the debtholders. If either of these assumptions is relaxed, debt automatically becomes subject to the risk of default. In the sequel, stockholders are assumed to have limited liability. This means that the firm's cash flow is the only source out of which the obligations to the debtholders can be met. Consequently, if the obligations to the debtholders exceed the firm's cash flow, the firm defaults and is declared bankrupt. This makes the bankruptcy condition, $b$, in single-period models:

$$(4.5) \qquad b = R > \tilde{x}$$

If $b$ obtains[5], stockholders are protected by limited liability and receive nothing. So, the end of period value of equity, $Y_e$, is:

$$\begin{aligned} Y_e &= 0 && \text{if } \tilde{x} < b \ , \\ &= (1-\tau)(\tilde{x}-R) && \text{if } \tilde{x} \geq b \ (b=R). \end{aligned}$$

The equilibrium value of equity, $V_e$, is the present value of the expectation of $Y_e$, which equals:

$$(4.6) \qquad V_e = \frac{(1-\tau)\int_b^{\infty}(\tilde{x}-R)f(\tilde{x})d\tilde{x}}{(1+r)}$$

If stockholders are protected by limited liability, it is reasonable that debtholders are granted the same protection. Limited liability for debtholders means that, if the bankruptcy condition obtains, they only receive the cash flow if it is positive; they are not required to accept a negative cash flow. In the literature, frequently no distinction is made between a positive and a negative cash flow[6]. This means that in those cases it is implicitly assumed either that the debtholders bear the loss in case of a negative cash flow or that the cash flow is always positive[7]. On the other hand, the assumption of limited liability for all security holders means that a negative cash flow is included in neither the value of equity nor the value of debt and, hence, constitutes a "leak" in the model[8]. Under the assumption of limited liability for all security holders the end of period value of debt is:

$$\begin{aligned} Y_d &= 0 && \text{if } 0 \geq \tilde{x} \ , \\ &= \tilde{x} && \text{if } 0 < \tilde{x} < b \ , \\ &= R && \text{if } b \leq \tilde{x} \ (b=R) \ . \end{aligned}$$

-------

[5] *Note that bankruptcy is defined more sharply here than it was in Chapter* 3. *Bankruptcy refers to all states of nature in which* (4.5) *obtains.*

[6] *For example by Kim* (1978), *Jonkhart* (1980) *and Chen and Kim* (1979).

[7] *A strictly positive cash flow is, of course, incompatible with the use of an open ended distribution for* $\tilde{x}$, *such as the normal distribution.*

[8] *One may think of this leak as the losses borne by e.g. employees, government and social security institutions.*

The equilibrium value of debt is the present value of the expectation of $Y_d$:

$$V_d = \frac{\int_0^b \tilde{X} f(\tilde{X})d\tilde{X} + \int_b^\infty R f(\tilde{X})d\tilde{X}}{(1+r)}$$

Using the following definition of the probability of bankruptcy $F$:

$F = \int_{-\infty}^b f(\tilde{X})d\tilde{X}$ , this can be written as :

$$(4.7) \quad V_d = \frac{\int_0^b \tilde{X} f(\tilde{X})d\tilde{X} + R(1-F)}{(1+r)} .$$

Adding up (4.6) and (4.7) and rearranging terms produces the following expression for the value of the firm:

$$(4.8) \quad V = \frac{\int_0^\infty \tilde{X} f(\tilde{X})d\tilde{X} - \tau \int_b^\infty \tilde{X} f(\tilde{X})d\tilde{X} + \tau R(1-F)}{(1+r)} .$$

The optimal capital structure and debt capacity are found by differentiating (4.8) and (4.7) with respect to $R$:

$$(4.9) \quad \frac{\partial V}{\partial R} = \frac{\tau(1-F)}{(1+r)} > 0 \ ;$$

$$(4.10) \quad \frac{\partial V_d}{\partial R} = \frac{(1-F)}{(1+r)} > 0 \ .$$

Compared with (4.4) and (4.5), the term $(1-F)$, i.e. the probability that the firm will not go bankrupt, is added in (4.9) and (4.10). So, while (4.4) represents the present value of the tax savings associated with a marginal amount of debt financing, (4.9) represents the present value of of the expected marginal tax savings. In a similar way, (4.5) and (4.10) represent the marginal benefits for the value of debt. Since (4.9) and (4.10) only contain benefits, the conclusions of optimal capital structure and corporate debt capacity at a 100% debt financing remain unchanged. This, of course, repeats the well known conclusion that the introduction of risky debt alone does not cause an optimal capital structure at below a 100% debt financing[9]. To ensure an "internal" optimal capital structure a cost of debt financing is needed.

-------

[9] *This result was first proved within a state-preference framework by Stiglitz* (1969) *and within the mean-variance approach by Rubinstein* (1973)

### 4.3.3 *The effect of bankruptcy costs*

The specification of bankruptcy costs

The results of the analyses so far are unsatisfactory in that the conclusions of optimal capital structure and corporate debt capacity at a 100% debt financing are poor predictors of firm behaviour in practice. The cause of these extreme implications is, of course, that the only market imperfection introduced so far, i.e. taxes, is beneficial for debt financing. To mitigate the extreme implications, a counterbalancing cost of debt financing is needed and the cost that was first suggested (among others by Robichek and Myers, 1965, and Baxter, 1967) were bankruptcy costs. Since then, the nature and magnitude of bankruptcy costs have extensively been discussed in the literature[10], and rigorous analyses of their influence on the financing decision have been provided[11].

In models of optimal capital structure, bankruptcy costs can be specified in several ways: in an implicit way (as is done by e.g. Scott, 1976), or in an explicit way and, in the latter case, as a function of the cash flow (as is done by e.g. Kim, 1978) or as a function of the payments promised to the debtholders (as is done by e.g. Castanias, 1983). Implicit bankruptcy costs occur when, upon bankruptcy, the firm's assets are sold in imperfect secondary markets at a price which is less than the market value of a well managed non-bankrupt firm. This specification is primarily suitable for multi-period models, in order to distinguish between the, possibly perpetual, cash flows in case of continuity and the one terminal cash flow in case of bankruptcy. In single-period models, the case of continuity is not accounted for, so there is always (i.e. bankrupt or not) only one cash flow, which makes an implicit specification less obvious. In addition to the foregoing, it should be noted that the implicit specification of bankruptcy costs requires the firm to be liquidated (i.e. its assets dismantled and sold). Thus, it necessitates stronger assumptions than bankruptcy, which is defined as occurring if the firm fails to meet its obligations to its debtholders. When the bankruptcy condition obtains, the ownership of the firm is transferred from the stockholders to the debtholders. This transfer may be

-------

[10] *See e.g. Kim* (1978), *Chen and Kim* (1979) *and Jonkhart* (1980).

[11] *Rigorous analyses are provided by, among others, Kraus and Litzenberger* (1973), *who use a state-preference framework, Scott* (1976), *who uses a mean-variance approach in a multi-period setting and Kim* (1978), *who uses the same approach in a single-period setting.*

costless or costly and, after the transfer, the debtholders may liquidate the firm or not, whatever is in their best interest. In fact, stockholders should liquidate the firm before bankruptcy occurs, if the liquidation value of its assets exceeds the firm's market value.

In the present single-period context, the analysis is limited to an explicit specification of bankruptcy costs, viz. bankruptcy costs as a function of the cash flow. The incorporation of bankruptcy costs as a function of the payments to the debtholders into the model developed in the previous subsection can be shown to produce indeterminate results (Van der Wijst, 1986). In addition to this, it can be noted that it seems more reasonable to relate bankruptcy costs to the amount actually transferred in case of bankruptcy than to the amount promised to transfer in case the firm is not bankrupt. Therefore, the case of bankruptcy costs as a function of the payments promised to the debtholders is not further elaborated here. An example of such an elaboration can be found in Castanias (1983)[12].

The effect of bankruptcy costs

The introduction of the bankruptcy costs into the model developed in the previous subsection does not change the valuation of equity. In case of bankruptcy, stockholders are protected by limited liability and receive nothing, and if the firm is not bankrupt, there are, of course, no bankruptcy costs. So the valuation formula for equity remains as in (4.6):

$$(4.6)\qquad V_e = \frac{(1-\tau)\int_b^\infty (\tilde{x}-R)f(\tilde{x})d\tilde{x}}{(1+r)}$$

The valuation of debt changes with the amount of bankruptcy costs. If, at the end of the period, the cash flow is insufficient to meet the obligations to the debtholders, the firm is transferred to them. There are costs associated with this transfer. Debtholders receive the positive balance, if any, of the cash flow and these costs:

$$\begin{aligned} Y_d &= 0 && \textit{if } \tilde{x}<0 , \\ &= \tilde{x}-B(\tilde{x}) && \textit{if } 0\leq\tilde{x}<b , \\ &= R && \textit{if } \tilde{x}\geq b \ (b=R) , \end{aligned}$$

-------

[12] *Castanias avoids indeterminate results by assuming unlimited liability, although the consequence of riskless debt is omitted. See Van der Wijst* (1986, *p.* 22) *for a more detailed discussion.*

where $B(\tilde{x})$ represents bankruptcy costs as a function of $\tilde{x}$ (see also table 4.1A ). The equilibrium value of debt is given by:

$$V_d = \frac{\int_0^b (\tilde{x} - B(\tilde{x}))f(\tilde{x})\, d\tilde{x} + \int_b^\infty Rf(\tilde{x})d\tilde{x}}{(1+r)}$$, which yields:

$$(4.11) \quad V_d = \frac{\int_0^b (\tilde{x} - B(\tilde{x}))f(\tilde{x})d\tilde{x} + R(1-F)}{(1+r)} .$$

The value of the firm is the sum of (4.6) and (4.11):

$$(4.12) \quad V = \frac{\int_0^\infty \tilde{x}f(\tilde{x})d\tilde{x} - \tau\int_b^\infty \tilde{x}f(\tilde{x})d\tilde{x} - \int_0^b B(\tilde{x})f(\tilde{x})d\tilde{x} + \tau R(1-F)}{(1+r)} .$$

The firm's optimal capital structure and debt capacity are found by differentiating (4.12) and (4.11) with respect to $R$, which produces:

$$(4.13) \quad \frac{\partial V}{\partial R} = \frac{\tau(1-F) - B(R)f(R)}{(1+r)} ;$$

$$(4.14) \quad \frac{\partial V_d}{\partial R} = \frac{(1-F) - B(R)f(R)}{(1+r)} .$$

Optimal capital structure is reached if (4.13) equals zero, i.e. if:

$$(4.15) \quad \frac{\tau(1-F)}{(1+r)} = \frac{B(R)f(R)}{(1+r)} .$$

The left hand side of (4.15) represents the present value of the marginal expected tax savings, and the right hand side represents the present value of the marginal bankruptcy costs. So, optimal capital structure is reached as the marginal benefits of debt financing equal the marginal costs. The same result applies to corporate debt capacity, on the understanding that, to the debtholders, the benefits of debt financing are the payments to them and not the tax savings associated with these payments:

$$(4.16) \quad \frac{(1-F)}{(1+r)} = \frac{B(R)f(R)}{(1+r)} .$$

Because the marginal bankruptcy costs are positive (by assumption), optimal capital structure and corporate debt capacity are reached before bankruptcy is certain (i.e. $F<1$). From a comparison of (4.15) and (4.16), it can be seen that as long as $0<\tau<1$, optimal capital structure is reached before corporate debt capacity. It can be shown that for normally distributed cash flows the second order conditions for (4.13) and (4.14) are satisfied.

Comparative statics analysis

Since the first and second order conditions for (4.13) and (4.14) are satisfied, the following comparative statics can be calculated[13]:

$$(4.17) \quad \frac{\partial^2 V}{\partial \tau \partial R} = \frac{(1-F)}{(1+r)} > 0 \, .$$

An increase in the tax rate increases the tax advantage associated with debt financing and thus raises the amount of debt in the optimal capital structure. Next, the probability of bankruptcy $F$:

$$(4.18) \quad \frac{\partial^2 V}{\partial F \partial R} = \frac{1}{(1+r)} \left( -\tau - \frac{\partial f(R)}{\partial F} B(R) \right) ,$$

$$(4.19) \quad \frac{\partial^2 V_d}{\partial F \partial R} = \frac{1}{(1+r)} \left( -1 - \frac{\partial f(R)}{\partial F} B(R) \right) ,$$

An increase in $F$ means an unambiguous density increase over the default range. Such an increase may occur as a result of a shift in the distribution of $\tilde{x}$ that decreases its expectation, but leaves the shape of the distribution unchanged. Since, for normal distributions, $\partial f(R)/\partial F$ can be positive as well as negative, it can be seen from (4.18) and (4.19) that a density increase over the default range has ambiguous effects on optimal capital structure and debt capacity. The marginal benefits of debt financing decrease with increasing $F$, but the marginal bankruptcy costs increase or decrease, making

-------

[13] *Strictly speaking, a comparative static of the model is obtained by totally differentiating the first order condition with respect to R and some exogenous parameter, say E. If the first order condition $f(R,E) = 0$ for some R (say $R^*$), the comparative static is obtained by:*

$$\frac{\partial f}{\partial R} dR + \frac{\partial f}{\partial E} dE = 0, \text{ or } \left. \frac{dR}{dE} \right|_{R=R^*} = - \frac{\partial f/\partial E}{\partial f/\partial R}$$

*Because $\partial f/\partial R$ is the second order condition, which is required to be negative if comparative statics are to be calculated, the sign of the comparative static is determined by the "cross-partial" $\partial f/\partial E$ ($f$ is the first order condition $\partial . /\partial R$). Because we are primarily interested in the sign of the comparative static, the comparative statics analyses are confined to these cross-partials (see Silberberg, 1981, for a comprehensive discussion of the methodology of comparative statics).*

the result ambiguous. In a similar way, a mean preserving increase in the variance of $\tilde{x}$ has an ambiguous influence on both $F$ and $f(R)$ for normal distributions, which means an ambiguous influence on optimal capital structure and debt capacity.

The influence of a change in the marginal bankruptcy costs on optimal capital structure and debt capacity is given by:

$$(4.20)\quad \frac{\partial^2 V}{\partial B(R)\partial R} = \frac{-f(R)}{(1+r)} < 0,$$

$$(4.21)\quad \frac{\partial^2 V_d}{\partial B(R)\partial R} = \frac{-f(R)}{(1+r)} < 0$$

An increase in the marginal bankruptcy costs makes debt financing less attractive on the margin and, thus, lowers the firm's debt capacity as well as the amount of debt in the firm's optimal capital structure.

Finally, the influence of the risk free rate of return is:

$$(4.22)\quad \frac{\partial^2 V}{\partial r\partial R} = \frac{-(\tau(1-F) - B(R)f(R))}{(1+r)^2} = 0 ,$$

$$(4.23)\quad \frac{\partial^2 V_d}{\partial r\partial R} = \frac{-((1-F) - B(R)f(R))}{(1+r)^2} = 0 .$$

Since the risk free rate of return equally influences the benefits and costs of debt financing, it does not influence optimal or maximal leverage.

To conclude this subsection, a final remark is devoted to Kim's (1978) analysis, a part of which (especially section IV) closely resembles the analyses in this subsection. Kim assumes that: "$B(\tilde{x})$ is an implicit function of $\tilde{x}$, but no greater than $\tilde{x}$, i.e. $B(\tilde{x}) \leq \tilde{x}$. This upper limit on $B(\tilde{x})$ provides debtholders with limited liability in a nominal asset case in which the proceeds from liquidation are entirely consumed in administrative expenses with no distribution to debtholders" (page 50). However, as was already argued in section 3.2, this assumption does not provide debtholders with limited liability, nor does it exclude the possibility of negative bankruptcy costs, unless $\tilde{x}$ is assumed to be strictly positive. This latter assumption is, of course, incompatible with use of an open-ended distribution, like the normal one, as is done by Kim.

### 4.3.4 *The effect of agency costs*

Although the concept of agency costs is broadly applicable, we shall confine ourselves in this study to the agency costs of debt financing. The incorporation of various types of agency costs would result in the incorporation of a number of only slightly different terms, without altering the qualitative results of the analysis. In models of optimal capital structure, agency costs are expressed in almost the same way as bankruptcy costs[14]. A difference between the two, however, is that the former are only made in case of bankruptcy, while the latter are incurred regardless of the state of nature. In the present single-period context, agency costs can be looked upon as being made at the beginning of the period (e.g. bonding costs) or during the period (monitoring costs) but in any case before the firm's financial outcome becomes known. It is therefore not in accordance with the nature of agency costs to define a broad category of leverage related costs, including agency and bankruptcy costs, that are only incurred in case of bankruptcy (as is done by Bradley, Jarrell and Kim, 1984 and implicitly by Jonkhart, 1980).

Because only the agency costs of debt are considered here, the valuation formula for equity is unaffected by it and remains as in (4.6):

$$(4.6)\quad V_e = \frac{(1-\tau)\int_b^{\infty}(\tilde{x}-R)f(\tilde{x})d\tilde{x}}{(1+r)}$$

The value of debt is, of course, affected by agency costs: in costless capital markets, rational investors anticipate these costs and deduct them from the amount they would otherwise be willing to pay. Agency costs are usually assumed to be an increasing function of the amount of debt in the firm's capital structure (cf. Barnea, Haugen and Senbet, 1981). In this study, agency costs are specified as an increasing function of the payments promised to the debtholders. The above argument can be formally summarized as follows:

$$\begin{aligned} Y_d &= -A(R) && \text{if } \tilde{x}<0 \\ &= \tilde{x}-B(\tilde{x})-A(R) && \text{if } 0\leq\tilde{x}<R \\ &= R-A(R) && \text{if } R\leq\tilde{x}\ (=b) \end{aligned}$$

-------

[14]*Jensen and Meckling* (1976), *p.* 342, *consider bankruptcy costs an element of the agency costs associated with debt.*

where $A(R)$ represents agency costs as a function of $R$ (see also table 4.1A in the appendix to this chapter). The equilibrium value of debt is:

$$V_d = \left[\int_{\infty}^{0} -A(R)f(\tilde{x})d\tilde{x} + \int_{0}^{b}(\tilde{x}-B(\tilde{x})-A(R))f(\tilde{x})d\tilde{x} + \int_{b}^{\infty}(R-A(R))f(\tilde{x})d\tilde{x}\right]\frac{1}{(1+r)},$$

$$(4.28)\quad V_d = \frac{-A(R) + \int_0^b \tilde{x}f(\tilde{x})d\tilde{x} - \int_0^b B(\tilde{x})f(\tilde{x})d\tilde{x} + R(1-F)}{(1+r)}.$$

The value of the firm is the sum of (6) and (28).

$$(4.29)\quad V = \frac{\int_0^{\infty}\tilde{x}f(\tilde{x})d\tilde{x} - \tau\int_b^{\infty}\tilde{x}f(\tilde{x})d\tilde{x} - \int_0^b B(\tilde{x})f(\tilde{x})d\tilde{x} + \tau R(1-F) - A(R)}{(1+r)}.$$

Comparing (4.28) and (4.29) with (4.11) and (4.12) respectively, it can be seen that the only difference between these two sets of equations is the agency costs term $A(R)$. So, to obtain the first order conditions for optimal capital structure and debt capacity first derivative of $A(R)$, which is defined as $A'(R)$, is added to (4.13) and (4.14):

$$(4.30)\quad \frac{\partial V}{\partial R} = \frac{\tau(1-F) - B(R)f(R) - A'(R)}{(1+r)},$$

$$(4.31)\quad \frac{\partial V_d}{\partial R} = \frac{(1-F) - B(R)f(R) - A'(R)}{(1+r)}.$$

Optimal capital structure is reached if the present value of the expected marginal tax savings equals the present value of the expected marginal bankruptcy costs, plus the marginal agency costs. A similar conclusion applies to debt capacity. The second order conditions are satisfied if the second derivative of the agency costs function is $\geq 0$ (as is assumed in table 4.1A in the appendix). This means that the comparative statics of this model are as in (4.17) through (4.23). They are not repeated here. The only additional cross-partial that can be calculated is the influence of marginal agency costs on optimal capital structure and debt capacity:

$$(4.32)\quad \frac{\partial^2 V}{\partial A'(R)\partial R} = \frac{\partial^2 V_d}{\partial A'(R)\partial R} = \frac{-1}{(1+r)} < 0.$$

An increase in the marginal agency costs leads to less debt in optimal capital structure and to a lower debt capacity.

The foregoing shows the effect on the financing decision of two types of leverage related costs in combination with risky debt and limited liability. From this it becomes clear, that the number of leverage related costs (or leverage related benefits, for that matter) can be freely extended without consequences for the qualitative results. For instance, a term representing the agency costs of equity may be included, inversely related to the payments promised to the debtholders. Or the bankruptcy costs function may be split into two parts, each representing different aspects of the costs of financial distress. These extensions will doubtlessly enhance the degree of realism of the model (as well as the degree of complexity). However, the qualitative results of the model (which constitute the basis of empirical testing), will remain unaltered by these extensions, which are, therefore, not further elaborated.

### 4.3.5. *The effect of informational differences*

In this subsection, the first assumption made in section 4.3.1, i.e. the assumption that investors' expectations are homogeneous, is relaxed. Investors may now have different expectations, for instance based on differences in information. The essence of these informational differences is that there is no longer one distribution of the firm's cash flows, but that different market participants have different expectations regarding this cash flow. These expectations can be influenced by all sorts of information (as the every day practice of stock exchanges shows). In the sequel of this subsection, it is assumed that the debtholders' expectations regarding the firm's cash flow differ from the owners' expectations. Between them, debtholders and owners are assumed to share the same expectations. This means that the firm may have a different value for debtholders than it does for owners, which is incompatible with the single price law of markets. Therefore, it is assumed that the market for debt is strictly separated from the market for equity, excluding the possibility of arbitrage through home-made leverage. Although, at first glance, this assumption may seem restrictive and unrealistic, it should be noted that the most important suppliers of debt, i.e. the commercial banks, are, in many countries, not allowed to supply equity capital[15].

-------

[15] *For instance, prior to* 1981 *banks in The Netherlands were practically not allowed to supply equity to firms, in order to avoid conflicts of interest between the banks and their clients. In* 1981, *an alteration of the law enabled banks to start so-called private participation companies, which are partly government guaranteed* (*see also section* 2.4).

The above argument can be formalized as follows. Assuming that the cash flow according to the owners is $\tilde{x}$ with density function $f_x(\tilde{x})$, and the cash flow according to the debtholders is $\tilde{w}$, with density function $f_w(\tilde{w})$, such that $f_x(\tilde{x}) \neq f_w(\tilde{w})$, the valuation formulae become:

$$Y_e = 0 \qquad \text{if } \tilde{x} < R ,$$

$$= (1-\tau)(\tilde{x}-R) \qquad \text{if } \tilde{x} \geq R .$$

$$(4.33) \quad V_e = \frac{(1-\tau)\int_R^\infty (\tilde{x}-R) f_x(\tilde{x}) d\tilde{x}}{(1+r)} ,$$

Which is equivalent to (4.6).

$$Y_d = -A(R) \qquad \text{if } 0 \geq \tilde{w} ,$$

$$= \tilde{w} - B(\tilde{w}) - A(R) \qquad \text{if } 0 < \tilde{w} < R ,$$

$$= R - A(R) \qquad \text{if } \tilde{w} \geq R .$$

where $B(\tilde{w})$ represents the bankruptcy costs as a function of $\tilde{w}$.

$$(4.34) \quad V_d = \frac{-A(R) + \int_0^R (\tilde{w} - B(\tilde{w})) f_w(\tilde{w}) d\tilde{w} + R(1-F_w)}{(1+r)} ,$$

where $F_w = \int_{-\infty}^R f_w(\tilde{w}) d\tilde{w}$, i.e. the probability of bankruptcy according to the debtholders.

Similarly $F_x = \int_{-\infty}^R f_x(\tilde{x}) d\tilde{x}$. Since $V = V_e + V_d$ :

$$(4.35) \quad V = \frac{(1-\tau)\int_R^\infty (\tilde{x}-R) f_x(\tilde{x}) d\tilde{x} + \int_0^R (\tilde{w} - B(\tilde{w})) f_w(\tilde{w}) d\tilde{w} + R(1-F_w) - A(R)}{(1+r)} .$$

Since $f_x(\tilde{x}) \neq f_w(\tilde{w})$, (4.35) cannot be simplified and the first order conditions for the value of equity and the value of debt have to be calculated separately.

$$\frac{\partial V}{\partial R} = \frac{\partial V_d}{\partial R} + \frac{\partial V_e}{\partial R} .$$

The first order conditions are:

$$\text{(4.36)} \quad \frac{\partial V_e}{\partial R} = \frac{-(1-F_x)(1-\tau)}{(1+r)} < 0 \, ,$$

$$\text{(4.37)} \quad \frac{\partial V_d}{\partial R} = \frac{(1-F_w) - B(R)f_w(R) - A'(R)}{(1+r)} \, .$$

The interpretation of (4.37) is analogous to the interpretation of (4.31). Since (4.36) is strictly negative, it can be seen (once more) that the firm's optimal capital structure, being the sum of (4.36) and (4.37), is reached before the firm's debt capacity and before bankruptcy is certain.

It can be shown that, for normally distributed cash flows, the second order condition for debt capacity is satisfied and the second order condition for optimal capital structure is satisfied if the following two conditions hold:

$$\frac{f_x(R) - f_w(R)}{f_x(R)} \leq \tau \, , \qquad \text{and} \qquad \frac{F_x - F_w}{(1-F_x)} \leq -\tau \, .$$

These conditions refer to the relation between $f_w(\tilde{w})$ and $f_x(\tilde{x})$, as could be expected. If we assume that these conditions are satisfied, the following comparative statics can be calculated.

$$\text{(4.38)} \quad \frac{\partial^2 V}{\partial \tau \partial R} = \frac{\partial^2 V_e}{\partial \tau \partial R} = \frac{(1-F)}{(1+r)} > 0 \, ,$$

$$\text{(4.39)} \quad \frac{\partial^2 V}{\partial F_x \partial R} = \frac{\partial^2 V_e}{\partial F_x \partial R} = \frac{(1-\tau)}{(1+r)} > 0 \, ,$$

$$\text{(4.40)} \quad \frac{\partial^2 V}{\partial F_w \partial R} = \frac{\partial^2 V_d}{\partial F_w \partial R} = \frac{-1 - B(R)\partial f_w(R)/\partial F_w}{(1+r)} \gtrless 0 \, ,$$

$$\text{(4.41)} \quad \frac{\partial^2 V}{\partial B(R) \partial R} = \frac{\partial^2 V_d}{\partial B(R) \partial R} = \frac{-f_w(R)}{(1+r)} < 0 \, ,$$

$$\text{(4.42)} \quad \frac{\partial^2 V}{\partial A'(R) \partial R} = \frac{\partial^2 V_d}{\partial A'(R) \partial R} = \frac{-1}{(1+r)} < 0 \, ,$$

$$(4.43)\quad \frac{\partial^2 V}{\partial r \partial R} = \frac{\partial^2 V_e}{\partial r \partial R} + \frac{\partial^2 V_d}{\partial r \partial R} = \frac{(1-\tau)(1-F_x)}{(1+r)^2} - \left[\frac{(1-F_w)-B(R)f_w(R)-A'R}{(1+r)^2}\right] = 0 .$$

These comparative statics (or cross-partials) are essentially the same as the ones derived in the previous subsections and need no further elucidation.

## 4.4 Multi-period models

### 4.4.1 *Introduction: additional assumptions and redefinitions*

In the literature, multi-period models occur far less frequently than single-period models. An obvious reason for this is the computational complexity of modelling the valuation of corporate claims in a multi-period context. Another reason may be that the results of single-period models can be interpreted in multi-period terms, as is already mentioned in section 4.2. However, a thorough treatment of the multi-period leverage decision requires explicit multi-period modelling. For this modelling, some additional assumptions have to be made and some variables have to be redefined. The first assumption is that of a perpetual model, i.e. the firm can exist indefinitely as long as it does not go bankrupt. This means that the cash flow ($\tilde{x}$) and the payments to the debtholders ($R$) are redefined as perpetuities (cf. Haley and Schall, 1979, p. 299). Further assumptions regarding $\tilde{x}$ and $R$ may affect the bankruptcy condition and are therefore made on a preliminary basis. With this restriction, the cash flow is assumed to have an intertemporally independent and identical probability distribution in each future period. The payments to the debtholders only comprise interest payments (no redemption of the principal sum) and are assumed to be fixed over time. The corporate tax rate ($\tau$) and the risk free rate of return ($r$) are also assumed to be fixed over time. Note that these assumptions imply that "every thing is constant over time", so that the multi-period elements are kept at their minimum.

The multi-period redefinition of the bankruptcy condition requires a separate discussion. In the previous section, bankruptcy was assumed to occur whenever the contractual obligations to the debtholders exceed the firm's cash flow (i.e. when $R>\tilde{x}$). In multi-period models, this condition is also necessary but not sufficient. The multi-period context requires the value of future income streams to be taken into account. For instance, the firm's owners would lose a possibly perpetual income stream if the firm would go bankrupt (i.e. be transferred to the debtholders) in case of a -possibly very small- loss in only one period[16]. This means that the value of future income streams has to be accounted for in the bankruptcy conditions to prevent the firm from going bankrupt when this value exceeds the loss in a single-period. It also means that if in a given period a loss occurs which is not large enough to cause bankruptcy, funds to the amount of the loss have to be raised or the firm will as yet go bankrupt. In principle, the firm can raise funds by selling equity, debt or (physical) assets. In each of these cases, the limit on the amount that can be raised is determined by the value of these three items. This value is to be incorporated into the bankruptcy condition to account for the value in future periods. The possibilities to sell debt or physical assets may be restricted by covenants in the debt contract.

Note that, in the absence of these covenants, it may be optimal for the owners to expropriate debtholders by issuing new debt, the claims of which are not subordinate to those of the senior debt. For the latter, the contractual interest payments remain the same, while the risk of default increases because of the junior debt claims. As a result, the existing debt will decrease in value. This is also true for risk-neutral investors. Owners' wealth will be maximal when the expropriation effect is maximal. Because expropriation means the transfer of wealth from (senior) debtholders to owners, maximizing the wealth of the owners is no longer equivalent to maximizing the sum of the values of debt and equity (i.e. maximizing firm value). In a similar way, the owners may sell the firm's productive assets, distribute the proceeds among them and leave the debtholders with an "empty" firm (see Haley and Schall, 1979, p. 478-479 for a more general discussion). The crucial assumption for these expropriations is, of course, that they were not anticipated by the debtholders. However, expropriation is not a very fruitful way to approach the problem of optimal capital structure, because in the long run market participants are likely to find ways to protect themselves against it. Therefore it is usually assumed in models of optimal structure that debtholders are protected against unanticipated transfers of their wealth to the owners by restrictive covenants in the debt contract. In this way, the firm's optimal capital structure becomes a Pareto optimum with respect to all the firm's security holders. In this study, debtholders are also assumed to be protected against expropriation.

If expropriation is ruled out, the question of how funds are to be raised in case of a loss becomes easier to answer. As Scott (1976, p. 37-38)

-------

[16] *A similar situation is described by Stiglitz* (1974, *p.* 856-857).

argues, it is always optimal to sell equity because the protective covenants in the debt contract either prohibit the sale of debt and assets or require the debtholders to be compensated for the decrease in value of the outstanding debt that results from the sale of new debt or assets. In the latter case, the funds to be raised should not only cover the loss but also the compensation to the debtholders. No compensation is required if equity is sold. In this way, Scott arrives at the bankruptcy conditions that has also been suggested by Stiglitz (1974): bankruptcy occurs if the cash flow is insufficient to meet the obligations to the debtholders and the firm cannot issue enough equity to cover the loss. Clearly, the proceeds of an equity issue cannot exceed the current value of equity, so the bankruptcy condition obtains when $R>\tilde{x}+V_e$. If the loss, $(\tilde{x}-R)$, is assumed to induce a tax rebate of $-\tau(\tilde{x}-R)$, that can be used immediately to pay the interest bill, the bankruptcy condition becomes: $R>\tilde{x}+V_e-\tau(\tilde{x}-R)$, which can be written as $\tilde{x}<R-V_e/(1-\tau)$. This bankruptcy condition, which is also employed by Scott (1976), will be used in the sequel of this section[17]. Since the bankruptcy condition does not account for the sale of debt or assets, the assumptions regarding $\tilde{x}$ and $R$, that were made earlier on a preliminary basis, can now be made without restriction.

### 4.4.2 *The MM – position*

In the multi-period context, the MM-case is again presented as a starting point of the analysis. Following the same methodology as is used in previous sections, in each subsequent model one or more assumptions will be relaxed and the effects on the valuation of corporate claims will be analysed. The assumptions, on which the MM-position relies, are in the multi-period context essentially the same as in the single-period context. Investors are assumed to share the same subjective expectations with regard to future events. All market participants are assumed to have equal and costless access to all relevant information. Bankruptcy is assumed to be costless. Owners are assumed to have unlimited liability and enough personal wealth to meet the contractual obligations to the debtholders if these cannot be met out of the firm's cash flow. This means that debt is riskless and that the owners always receive the -positive or negative- cash flow after interest and taxes. The valuation formulae for this case are developed as follows. At the end of the period, the owners receive the cash flow after the interest and taxes. If we

-------

[17] *Note that the bankruptcy condition does not contain option elements other than those comprised in* $\tilde{x}$.

assume that profits and losses are fiscally treated in the same way, this amounts to $(1-\tau)(\tilde{x}-R)$. In addition to the returns over the period, the owners' wealth also comprises the value of all future income streams. Since all relevant variables ($r$, $R$, $\tau$ and the distribution of $\tilde{x}$) are assumed to be fixed over the time, the value of equity, $V_e$, represents the end of period value of all future income streams. So, the end of period value of equity, $Y_e$, equals:

$$Y_e = (1-\tau)(\tilde{x}-R)+V_e .$$

For risk neutral investors the equilibrium value of equity, $V_e$, is the present value, discounted at the risk free interest rate, of the expectation of $Y_e$:

$$V_e = \frac{E(Y_e)}{(1+r)} = \frac{(1-\tau)\int_{-\infty}^{+\infty}(\tilde{x}-R)f(\tilde{x})d\tilde{x} + V_e}{(1+r)}, \text{ which equals}$$

$$(4.44)\quad V_e = \frac{(1-\tau)(\bar{x}-R)}{r} .$$

Similarly, the end of period value of debt, $Y_d$, comprises the interest payments over the period, $R$, plus the value of all future interest payments. Since the latter are assumed to be fixed over time, their end of period value is given by $V_d$, the value of debt:

$$Y_d = R+V_d .$$

For risk neutral investors, the equilibrium value of debt, $V_d$, is the present value of the expectation of $Y_d$:

$$V_d = \frac{E(Y_d)}{(1+r)} = \frac{R+V_d}{(1+r)}, \text{ which leads to}$$

$$(4.45)\quad V_d = \frac{R}{r} .$$

Since $V = V_e + V_d$,

$$(4.46)\quad V = \frac{(1-\tau)\bar{x} + \tau R}{r} .$$

Comparing (4.44), (4.45) and (4.46) with (4.1), (4.2) and (4.3) respectively we see that only the denominator is changed to reflect the perpetual model. The MM conclusion that the value of the levered firm equals the value of the unlevered firm plus the tax advantage of debt remains unchanged. So, optimal capital structure and corporate debt capacity are reached at a hundred percent debt financing, as differentiating (4.46) and (4.45) with respect to $R$ shows:

$$\frac{\partial V}{\partial R} = \frac{\tau}{r} > 0, \text{ and} \tag{4.47}$$

$$\frac{\partial V_d}{\partial R} = \frac{1}{r} > 0\,. \tag{4.48}$$

These results are equivalent to the single-period results in (4.4) and (4.5).

### 4.4.3 *The effect of limited liability and the risk of default*

The concept of limited liability and its counterpart, the risk of default, have already been discussed in section 4.3.2 and need no further elaboration in the present multi-period context. Under the assumption that all security holders have limited liability, implying that debt is risky, valuation formulae for the firm's securities can be derived as follows. If the firm does not go bankrupt, the end of period value of equity comprises the dividends (cash flows after interest and taxes) over the period, and the value of the expected future dividends. The latter equals $V_e$, since all relevant variables ($r$, $R$, $\tau$ and the distribution of $\tilde{x}$) are assumed to be fixed over time. In case of bankruptcy, the owners are protected by limited liability and receive nothing. So the end of period value of equity, $Y_e$, is:

$$\begin{aligned} Y_e &= 0 && \text{if } \tilde{x} < b \\ &= (1-\tau)(\tilde{x}-R)+V_e && \text{if } \tilde{x} \geq b\,, \end{aligned}$$

in which $b=R-V_e/(1-\tau)$, the multi-period bankruptcy condition developed in section 4.4.1. Given risk neutrality, the equilibrium value of equity, $V_e$, is the present value of the expectation of $Y_e$:

$$V_e = \frac{E(Y_e)}{(1+r)} = \frac{(1-\tau)\int_b^\infty (\tilde{x}-R)\,f(\tilde{x})d\tilde{x} + V_e\int_b^\infty f(\tilde{x})d\tilde{x}}{(1+r)}\,.$$

If we define the probability of bankruptcy as $F=\int_{-\infty}^{b} f(\tilde{x})d\tilde{x}$, this can be written as:

$$(4.49)\quad V_e = \frac{(1-\tau)\int_b^\infty \tilde{x}f(\tilde{x})\,d\tilde{x} - (1-\tau)R(1-F)}{(r+F)}\,.$$

Compared with (4.6), the denominator of (4.49) is adapted to the perpetual model, while the numerator is the same as in the equivalent single-period model. Note that (4.49) defines $V_e$ only implicitly since $V_e$ is also an element of $b$, which is the lower limit of integration.

The valuation formula for debt can be derived along similar lines. If the firm does not go bankrupt, the end of period value of debt, $Y_d$, consists of the interest payment over the period, $R$, plus the value of all expected future interest payments. Since the latter are assumed to be constant over time, the value of the expected future interest payments is given by $V_d$. The composition of the value of debt in case of bankruptcy is somewhat more complex than when the firm is not bankrupt. If the bankruptcy condition obtains, the firm is transferred from the owners to the debtholders, and the assumption made in section 4.4.2, which is still in force, stipulates that this transfer is costless. The value to be transferred consists of the proceeds over the period an the value of future income streams, if any, while limited liability for the debtholders requires only positive values to be transferred.

At this point, some additional assumptions regarding the distribution of a bankrupt estate are necessary. It is possible to assume that debtholders have absolute priority over all other creditors so that they receive all positive values. However, it seems more realistic to assume that debtholders find themselves in a position comparable with that of the owners, i.e. that they have to trade-off the proceeds over the period against the value of future income streams. The proceeds over the period consist of the cash flow and the tax rebate and are –in view of the bankruptcy condition– likely to be negative. If this "loss" exceeds the value of expected future income streams, the firm will be abandoned so that the debtholders receive nothing. If the loss does not outweigh the future value, it will be covered. One may think of this situation as one in which the operations of a bankrupt firm can be continued once the claims of other creditors than the debtholders (e.g. employees and suppliers) are satisfied. Note that, since there are no transfer costs, it is not necessary for the debtholders to actually continue the firm's operations; they may also sell the firm.

So, it can be concluded that limited liability for debtholders in a multi-period context requires a "continuity condition" in addition to (and comparable with) the bankruptcy condition that ensures limited liability for the stockholders. This continuity condition specifies that the value to be transferred to the debtholders should be positive. The value to be transferred consists of the proceeds over the period (tax rebate and cash flow) plus the value of future income streams. Since all relevant variables ($r$, $R$, $\tau$ and the distribution of $\tilde{x}$) are assumed to be fixed over time, the value of future income streams is given by $V$, the value of the firm. This makes the "continuity condition" $\tilde{x}-\tau(\tilde{x}-R)+V>0$, which can be written as $\tilde{x}>(-V-\tau R)/(1-\tau)$. So the continuity condition, $c$, is $c=(-V-\tau R)/(1-\tau)$. Hence, the end of the period value of debt is:

$$
\begin{aligned}
Y_d &= R+V_d && \text{if} \quad \tilde{x}\geq b \\
&= \tilde{x}-\tau(\tilde{x}-R)+V && \text{if} \quad c\leq\tilde{x}<b \\
&= 0 && \text{if} \quad c>\tilde{x}
\end{aligned}
$$

In equilibrium, the value of debt, $V_d$, is:

$$V_d = \frac{E(Y_d)}{(1+r)} = \frac{\int_b^\infty (R+V_d)f(\tilde{x})d\tilde{x} + \int_c^b (\tilde{x}-\tau(\tilde{x}-R)+V)\, f(\tilde{x})d\tilde{x}}{(1+r)},$$

in which $c=(-V-\tau R)/(1-\tau)$. Using a new symbol for the probability of bankruptcy $F_b=\int_{-\infty}^b f(\tilde{x})d\tilde{x}$ and defining $F_c=\int_{-\infty}^c f(\tilde{x})d\tilde{x}$, the above expression can be written as:

$$(4.50) \quad V_d = \frac{R(1-F_b) + (1-\tau)\int_c^b \tilde{x}f(\tilde{x})d\tilde{x} + (\tau R+V_e)\int_c^b f(\tilde{x})d\tilde{x}}{(r+F_c)}.$$

The denominators in (4.50) and (4.49) are different and no simplification is reached by working out the terms of their sum. Hence, the value of the firm, $V=V_e+V_d$, is given by:

$$(4.51) \quad V = \frac{(1-\tau)\int_b^\infty (\tilde{x}-R)f(\tilde{x})d\tilde{x}}{(r+F_b)} + \frac{R(1-F_b) + (1-\tau)\int_c^b \tilde{x}f(\tilde{x})d\tilde{x} + (\tau R+V_e)\int_c^b f(\tilde{x})d\tilde{x}}{(r+F_c)}.$$

Differentiation of (4.49) and (4.50) with respect to $R$ to find the optimal capital structure and debt capacity produces the following results:

$$(4.52) \quad \frac{\partial V_e}{\partial R} = \frac{-(1-\tau)(1-F_b)}{(r+F_b)} < 0, \text{ and}$$

$$(4.53) \quad \frac{\partial V_d}{\partial R} = \frac{(1-F_b) + \int_c^b f(\tilde{x})d\tilde{x} \left[\tau - \frac{(1-\tau)(1-F_b)}{(r+F_b)}\right]}{(r+F_b)}$$

Equation (4.52) is strictly negative. The sign of (4.53) cannot be determined without further assumptions. Hence, the first order conditions for optimal capital structure and debt capacity may be satisfied. However, the second order conditions for optimal capital structure and debt capacity cannot be shown to hold[18]. As can be seen from (4.50) and (4.53), the complexity of the effects of an increase in $R$ on the component parts of the value of debt prohibits the on balance effect to be unambiguously determined without further assumptions. Hence, even in this early stage of the analyses the computational complexity of valuing corporate claims in a multi–period world prevents clear–cut results.

### 4.4.4 *The effect of bankruptcy costs*

In view of the results obtained in the previous subsection, the choice of the bankruptcy costs specification is limited. The addition of an explicit bankruptcy costs function to the formulae developed in section 4.4.3 would not alter the –indeterminate– qualitative results. So the obvious way to proceed with the analysis is to introduce implicit bankruptcy costs into the valuation formulae. Implicit bankruptcy costs were defined to occur when, upon bankruptcy, the firm's productive assets are sold in imperfect secondary markets at a price below the market value of an equivalent, well managed, non bankrupt firm. As is already noted in section 4.3.3, implicit bankruptcy costs require the firm to be liquidated. This is a stronger assumption than bankruptcy, which only requires the firm to be transferred to the debtholders. If the firm is liquidated upon bankruptcy, the case of continuity under the bankruptcy condition is not accounted for and, hence, the continuity condition is redundant. This simplifies the distribution of a bankrupt estate to a certain extent, because there are no future income streams to be traded–off against the loss in a period. Hence, the debtholders may be assumed to receive all positive values, while the negative values are not taken into account (see section 4.3.2 for a description of this "leak" in the model).

Neither the composition nor the distribution of a bankrupt estate affects the value of equity, which is zero under the bankruptcy condition. So

---

[18] *See Van der Wijst* (1986) *for a mathematical treatment of this matter.*

the value of equity remains as in the previous subsections:

$$(4.49)\quad V_e = \frac{(1-\tau)\int_b^{\infty} \tilde{x}f(\tilde{x})\,d\tilde{x} - (1-\tau)R(1-F)}{(r+F)} .$$

In the absence of a continuity condition the subscript of $F$ is redundant.

Using the concept of implicit bankruptcy costs, the valuation formula for debt is developed in the following manner. If the firm does not go bankrupt, the end of period value of debt, $Y_d$, consists of the interest payments over the period, $R$, and the value of the expected future interest payments, which can be represented by $V_d$. Upon bankruptcy, the firm is liquidated and its assets are sold in imperfect secondary markets, the proceeds of which are represented by $L$. In addition to the liquidation value there are also the proceeds over the period, i.e. the cash flow and the tax rebate. The latter is always positive but the former may take negative values. Limited liability for debtholders requires negative payments to be excluded, so the end of the period value of debt is:

$$\begin{aligned} Y_d &= R+V_d && \text{if } \tilde{x}\geq b \\ &= L-\tau(\tilde{x}-R)+\tilde{x} && \text{if } 0\leq\tilde{x}<b \\ &= L-\tau(\tilde{x}-R) && \text{if } \tilde{x}<b<0 \end{aligned}$$

in which $L$ is the liquidation value of the firm and $b$ is the by now familiar bankruptcy condition. For risk neutral investors, the equilibrium value of debt, $V_d$, is the present value of the expectation of $Y_d$:

$$V_d = \frac{\int_b^{\infty}(R+V_d)f(\tilde{x})d\tilde{x} + \int_0^{b}(L-\tau(\tilde{x}-R) + \tilde{x})\, f(\tilde{x})d\tilde{x} + \int_{-\infty}^{0}(L-\tau(\tilde{x}-R))\, f(\tilde{x})d\tilde{x}}{(1+r)} .$$

Rearranging terms we get:

$$(4.54)\quad V_d = \frac{R(1-F) + LF - \tau\int_{-\infty}^{b}\tilde{x}f(\tilde{x})\,d\tilde{x} + \tau RF + \int_0^{b}\tilde{x}f(\tilde{x})d\tilde{x}}{(r+F)} .$$

Note that (4.54) rests on the implicit assumption that the value of the bankrupt estate is insufficient to meet in full the debtholders' claims on the firm. If the value of the bankrupt estate exceeds the debtholders' claim, debt will be riskless and the owners may receive a liquidation payment, which would lead to a different set of valuation formulae. However, the case of fully secured (and hence riskless) debt is left out of consideration here.

The value of the firm, $V = V_e + V_d$, is the sum of (4.49) and (4.54):

$$(4.55)\quad V = \frac{\tau R - \tau\bar{x} + LF + \int_0^\infty \tilde{x} f(\tilde{x}) d\tilde{x}}{(r+F)} .$$

Note that both the tax–levy on the cash flow and the tax advantage of debt are certain, because of the tax–rebate in case of bankruptcy. Differentiating (4.55) and (4.54) with respect to $R$ to find the level of interest payments that maximizes the value of the firm and the value of debt respectively, we get:

$$(4.56)\quad \frac{\partial V}{\partial R} = \frac{\tau - \frac{(1+r)}{(r+F)} f(b)(V-L)}{(r+F)} ,$$

$$(4.57)\quad \frac{\partial V_d}{\partial R} = \frac{(1-F) + \tau F - \frac{(1+r)}{(r+F)} f(b)(V-L)}{(r+F)} .$$

Equation (4.56) can be interpreted as follows: $\tau$ represents the tax savings associated with a marginal amount of debt financing, while the second term can be thought of as the decrease in firm value due to a marginal increase in debt. The loss of value in case of bankruptcy is represented by $V-L$, while $((1+r)/(r+F))f(b)$, i.e. $\partial F/\partial R$, represents the marginal increase in the probability of bankruptcy. Equation (4.57) can be interpreted in a similar way. So the first order conditions in (4.56) and (4.57) are met if the marginal benefits of debt financing equal the marginal costs of debt financing. However, as was the case in the previous subsection, the second order conditions cannot be shown to hold[19].

So at this point the analysis wavers. The introduction of new variables in a model with indeterminate results is rather pointless. The possibility of an alternative (explicit) specification of bankruptcy costs is already excluded at the beginning of this subsection because of the results obtained in section 4.4.3. The only alternative is a different (simpler) specification of the valuation formulae. Such a specification is used by Scott (1976) and his analysis is the subject of the next subsection.

### 4.4.5 *The Scott model*

The model developed by Scott in his 1976 paper closely resembles the analysis

-------

[19] *See Van der Wijst* (1986) *for a mathematical treatment.*

in the previous subsection. The differences occur in the composition of a bankrupt estate. Scott assumes that the tax rebate is either lost upon bankruptcy or exhausted by the priority creditors[20]. A positive cash flow in case of bankruptcy is assumed to be used up in administrative costs. These assumptions do not affect the value of equity; the valuation formulae for equity in the Scott model is the same as the one developed in the previous subsection and is given by:

$$(4.49)\quad V_e = \frac{(1-\tau)\int_b^\infty \tilde{x} f(\tilde{x})\, d\tilde{x} - (1-\tau)R(1-F)}{(r+F)}.$$

The valuation formula for debt is somewhat simpler because the tax rebate and cash flow are assumed not to occur in case of bankruptcy. Scott also leaves the case of fully secured debt out of consideration, based on the argument that if debt is fully secured and the owners do not receive a liquidation payment, it is always optimal for the firm to issue some unsecured debt.
The end of period value of debt in the Scott model is:

$$\begin{aligned} Y_d &= R+V_d && \text{if } \tilde{x}\geq b \\ &= L && \text{if } \tilde{x}<b \end{aligned}$$

So the equilibrium value of debt is:

$$(4.58)\quad V_d = \frac{R(1-F) + LF}{(r+F)}.$$

The value of the firm, $V=V_e+V_d$, is given by:

$$(4.59)\quad V = \frac{(1-\tau)\int_b^\infty \tilde{x} f(\tilde{x})d\tilde{x} + \tau R(1-F) + LF}{(r+F)}.$$

Differentiating (4.59) and (4.58) with respect to $R$ we get the following first order conditions for optimal capital structure and corporate debt capacity:

---

[20] *Note that, although the tax rebate is not included in the bankrupt estate, it is still included in the bankruptcy condition used by Scott (as is explained in section* 4.4.1). *When the tax rebate would be eliminated from the bankruptcy condition, the Scott model would lose its "well behaved" character and produce indeterminate results (see Van der Wijst,* 1986, *p.* 60 −62, *for mathematical details).*

$$(4.60)\quad \frac{\partial V}{\partial R} = \frac{\tau(1-F) - \frac{(1+r)}{(r+F)}f(b)(R+V_d-L)}{(r+F)}, \text{ and}$$

$$(4.61)\quad \frac{\partial V_d}{\partial R} = \frac{(1-F) - \frac{(1+r)}{(r+F)}f(b)(R+V_d-L)}{(r+F)}.$$

It can be shown that the second order condition for optimal capital structure is satisfied for normally distributed cash flow if the following condition holds[21]:

$$(4.62)\quad \tau < \frac{(1+r)}{2(1-F)}.$$

The second order condition for debt capacity is satisfied for normally distributed cash flows if the following condition holds:

$$(4.63)\quad (1-F) < (r+F).$$

Condition (4.62) always holds if $\tau<.5$. If, for instance, $\tau=.6$ and $r=.1$, condition (4.62) holds if optimal capital structure is reached at a point where $F>.08$ and condition (4.63) holds if debt capacity is reached at a point where $F>.45$. If condition (4.62) and (4.63) hold, it can be seen from (4.60) and (4.61) that optimal capital structure is reached before corporate debt capacity and that both are reached before bankruptcy is certain. The following comparative statics can be calculated under the assumption that condition (4.62) and (4.63) hold:

$$(4.64)\quad \frac{\partial^2 V}{\partial\tau\partial R} = \frac{(1-F)}{(r+F)} > 0.$$

An increase in the tax rate increases the tax advantage of debt, which leads to more debt in the optimal capital structure. The effects of a shift in the probability of bankruptcy on optimal capital structure and debt capacity are:

$$(4.65)\quad \frac{\partial^2 V}{\partial F\partial R} = \frac{-\tau}{(r+F)} + \frac{(1+r)}{(r+F)^2}(R+V_d-L)\left[\frac{2f(b)}{(r+F)} - \frac{\partial f(b)}{\partial F}\right],$$

-------

[21] *See van der Wijst* (1986) *for an overview of the calculations, or Scott* (1976) *who uses a slightly different method.*

$$(4.66)\quad \frac{\partial^2 V_d}{\partial F \partial R} = \frac{-1}{(r+F)} + \frac{(1+r)}{(r+F)^2}(R+V_d-L)\left[\frac{2f(b)}{(r+F)} - \frac{\partial f(b)}{\partial F}\right].$$

The signs of (4.65) and (4.66) cannot be determined without further assumptions. The first term of (4.65) and (4.66) is strictly negative but the second term may be positive as well as negative ($\partial f(b)/\partial F$ may be positive or negative for normal distributions). As was noted in section 4.3.3, an increase in $F$ means an unambiguous density increase over the default range, which may occur as a result of a shift in the distribution of $\tilde{x}$ that decreases its expectation but leaves the shape of the distribution unaltered. Such a shift is seen from (4.65) and (4.66) to have ambiguous effects on optimal capital structure and debt capacity. Similarly, a mean preserving increase in the variance of $\tilde{x}$ has ambiguous effects on both $F(b)$ and $f(b)$ for normal distributions, and thus ambiguous effects on optimal capital structure and debt capacity.

Next, an increase in the firm's liquidation value has the following effect on optimal capital structure and debt capacity:

$$(4.67)\quad \frac{\partial^2 V}{\partial L \partial R} = \frac{\partial^2 V_d}{\partial L \partial R} = -\frac{(1+r)}{(r+F)^2}f(b)\left[\frac{\partial V_d}{\partial L} - \frac{\partial L}{\partial L}\right] = -\frac{(1+r)}{(r+F)^2}f(b)\left[\frac{F}{(r+F)} - 1\right]$$

$$= \frac{r(1+r)f(b)}{(r+F)^3} > 0 .$$

An increase in $L$ decreases the loss of value associated with bankruptcy $(R+V_d-L)$, which leads to more debt in the optimal capacity structure and to a higher debt capacity[22].

An increase in the risk free rate of return has the following effect:

$$(4.68)\quad \frac{\partial^2 V}{\partial r \partial R} = \frac{\partial^2 V_d}{\partial r \partial R} = \frac{(1-F)}{(r+F)^3}f(b)(R+V_d-L) + \frac{(1+r)}{(r+F)^3}f(b)V_d > 0 .$$

From (4.68) it can be seen that an increase in the risk free rate of return leads to more debt in the optimal capital structure[23] and to a higher debt capacity.

-------

[22] *Scott (1976, p. 47) presents a quantitatively different but qualitatively equivalent formula for* $\partial^2 V/\partial R \partial L$.

[23] *Scott (1976, p. 48) reports an ambiguous effect.*

### 4.4.6 *Some extensions of the Scott model*

Agency costs

Taking the Scott model presented in the previous subsection as a basis, the step–by–step procedure used so far can be continued by relaxing the assumption that there are no agency costs. Since the concept of agency costs is already discussed in section 4.3.4, this subsection is confined to the influence of agency costs in the Scott model. As in section 4.3.4, only the agency costs of debt are considered here. This means that the valuation formula for equity is unaffected by the introduction of agency costs and remains as in (4.49). Of course, the valuation of debt changes if agency costs
are introduced. Using the same specification for agency costs as in section 4.3.4, the end of period value of debt becomes:

$$Y_d = R+V_d-A(R) \qquad \text{if } \tilde{x}\geq b$$

$$= L-A(R) \qquad \text{if } \tilde{x}<b$$

in which $A(R)$ represents agency costs as a function of $R$ (cf. table 4.1A). In equilibrium, the value of debt is:

$$(4.69)\quad V_d = \frac{R(1-F) + LF - A(R)}{(r+F)}.$$

The value of the firm can be found by adding (4.49) and (4.69)

$$(4.70)\quad V = \frac{(1-\tau)\int_b^\infty \tilde{x}f(\tilde{x})d\tilde{x} + \tau R(1-F) + LF - A(R)}{(r+F)}.$$

The only difference between (4.69) and (4.70) on the one hand and (4.58) and (4.59) on the other hand is the agency costs term. So the first order conditions for optimal capital structure and debt capacity are obtained by adding the first derivative of $A(R)$ –defined as $A'(R)$– to (4.60) and (4.61)

$$(4.71)\quad \frac{\partial V}{\partial R} = \frac{\tau(1-F) - \frac{(1+r)}{(r+F)}f(b)(R+V_d-L) - A'(R)}{(r+F)}, \text{ and}$$

$$(4.72)\quad \frac{\partial V_d}{\partial R} = \frac{(1-F) - \frac{(1+r)}{(r+F)} f(b)(R+V_d-L) - A'(R)}{(r+F)}.$$

The second order conditions for (4.71) and (4.72) are satisfied if (4.62) and (4.63) hold and if the second derivative of the agency costs function equals or exceeds zero ($A''(R) \geq 0$, as is assumed in table 4.1A). Since agency costs are only a function of the payments promised to the debtholders, the comparative statics of this model are as in (4.64) through (4.68), while the influence of an increase in the marginal agency costs is:

$$(4.73)\quad \frac{\partial^2 V}{\partial A'(R)\partial R} = \frac{\partial^2 V_d}{\partial A'(R)\partial R} = \frac{-1}{(r+F)}.$$

An increase in the marginal agency costs makes debt financing less attractive on the margin, resulting in less debt in the optimal capital structure and a lower debt capacity.

Informational differences

Following the same procedure as we did in section 4.3.5, informational differences can be expressed by specifying different cash flows in the valuation formulae for debt and equity. The cash flow according to the owners is assumed to be $\tilde{x}$, with density function $f_x(\tilde{x})$, while the cash flow according to the debtholders is $\tilde{w}$, with density function $f_w(\tilde{w})$, such that $f_x(\tilde{x}) \neq f_w(\tilde{w})$.

This means that:

$$F_x = \int_{-\infty}^{b_x} f_x(\tilde{x})d\tilde{x} \quad \text{and} \quad F_w = \int_{-\infty}^{b_w} f_w(\tilde{w})d\tilde{w},$$

i.e. the probability of bankruptcy according to the owners and debtholders respectively. With these adaptations, the valuation formula for equity according to the owners is equivalent to (4.49):

$$(4.74)\quad V_e = \frac{(1-\tau)\int_{b_x}^{\infty}(\tilde{x}-R)f_x(\tilde{x})d\tilde{x}}{(r+F_x)},$$

in which $b_x$ is the bankruptcy condition according to the owners, i.e. $R-V_e/(1-\tau)$, where $V_e$ is given in (4.74). The bankruptcy condition according to the debtholders is $b_w=R-V_e^*/(1-\tau)$, where $V_e^*$ is the value of equity according to the debtholders:

$$(4.75)\quad V_e^* = \frac{(1-\tau)\int_{b_w}^{\infty} (\tilde{w}-R)f_w(\tilde{w})d\tilde{w}}{(r+F_w)} .$$

Similarly, the valuation formula for debt is equivalent to (4.69) with adapted $F$:

$$(4.76)\quad V_d = \frac{R(1-F_w) + LF_w - A(R)}{(r+F_w)} .$$

The value of the firm is the sum of (4.74) and (4.76):

$$(4.77)\quad V = \frac{(1-\tau)\int_{b_x}^{\infty} (\tilde{x}-R)f_x(\tilde{x})d\tilde{x}}{(r+F_x)} + \frac{R(1-F_w) + LF_w - A(R)}{(r+F_w)} .$$

No simplification is reached by working and terms of (4.77) and the first order conditions for the value of equity and the value of debt are calculated separately.

$$(4.78)\quad \frac{\partial V_e}{\partial R} = \frac{-(1-\tau)(1-F_x)}{(r+F_x)} < 0$$

$$(4.79)\quad \frac{\partial V_d}{\partial R} = \frac{(1-F_w) - \frac{(1+r)}{(r+F_w)} f_w(b_w)(R+V_d-L) - A'(R)}{(r+F_w)} .$$

Equations (4.78) and (4.79) are equivalent to (4.52) and (4.61) with adapted subscripts. The second derivatives of these valuation formulae are (see Van der Wijst, 1986, for mathematical details):

$$(4.80)\quad \frac{\partial^2 V_e}{\partial R^2} = \frac{(1-\tau)(1+r)^2 f_x(b_x)}{(r+F_x)^3} > 0$$

$$(4.81)\quad \frac{\partial^2 V_d}{\partial R^2} = \frac{1}{(r+F_w)} \left[ \frac{-2(1+r)}{(r+F_w)} + \frac{(1+r)^2 f_w(b_w)^2 (R+V_d-L)}{(r+F_w)^3} - \frac{(1+r)}{(r+F_w)} \frac{\partial f_w(b_w)}{\partial R}(R+V_d-L) - \frac{2(1+r)}{(r+F_w)} f_w(b_w) \frac{\partial V_d}{\partial R} - A''(R) \right] .$$

Equation (4.80) is positive. Equation (4.81) can be shown to be negative, but only for normally distributed cash flow and only if the equivalent of (4.63) obtains as well as the condition that $\partial V_d/\partial R = 0$. The latter condition is not satisfied in the firm's optimal capital structure, since $\partial V_e/\partial R < 0$. Hence, the second order condition for optimal capital structure cannot be shown to hold in general, whereas the second order condition for debt capacity holds under the three conditions mentioned above. Because of these (partly) indeterminate results, Scott–case is not further elaborated.

## 4.5 Conclusions

Reviewing the theoretical exercises in this chapter is likely to give rise to rather skeptical conclusions, as probably the most striking aspect of the exercises is the large number of models which produce indeterminate results. To illustrate, the step–by–step procedure used in the previous subsections is schematically represented in table 4.1. In this table, each step is represented by a block and the results of the models developed in each step are set at the bottom of the block. An "internal" solution for optimal capital structure and corporate debt capacity is denoted by MC=MR. If optimal capital structure and corporate debt capacity cannot be shown to exist, the results are labelled "indeterminate". References are placed at the top of blocks; the references can refer to either the original contribution to the theory or an elaboration of existing ideas in models.

As table 4.1 shows, there is no question of a tree–structured family of interrelated models. In fact, only one path in the table is practicable to the end, viz. the line of single–period models in which bankruptcy costs are specified as a function of the cash flow. The other four paths peter out in indeterminate results, often in an early stage of the analysis. So, modelling the valuation of corporate claims, even within the simple and rather restrictive framework of assumptions adopted in this study, quickly leads to models involving complex derivations. As more and more assumptions are relaxed, the results of the computations tend to become too complex to be evaluated, even in qualitative terms. The consequences of this complexity are not only that a comparatively large number of models produces indeterminate results, but also, and perhaps more seriously, that the models are sensitive for small (and by themselves trivial) changes in the specification of variables. For instance, the introduction of bankruptcy costs in single–period models produces an internal solution if these costs are specified as a function of the cash flow, but indeterminate results if these costs are considered a function of the payments to debtholders. In the multi–period setting, only the Scott–model and some of its extensions produce

table 4.1 schematic representation of the analyses

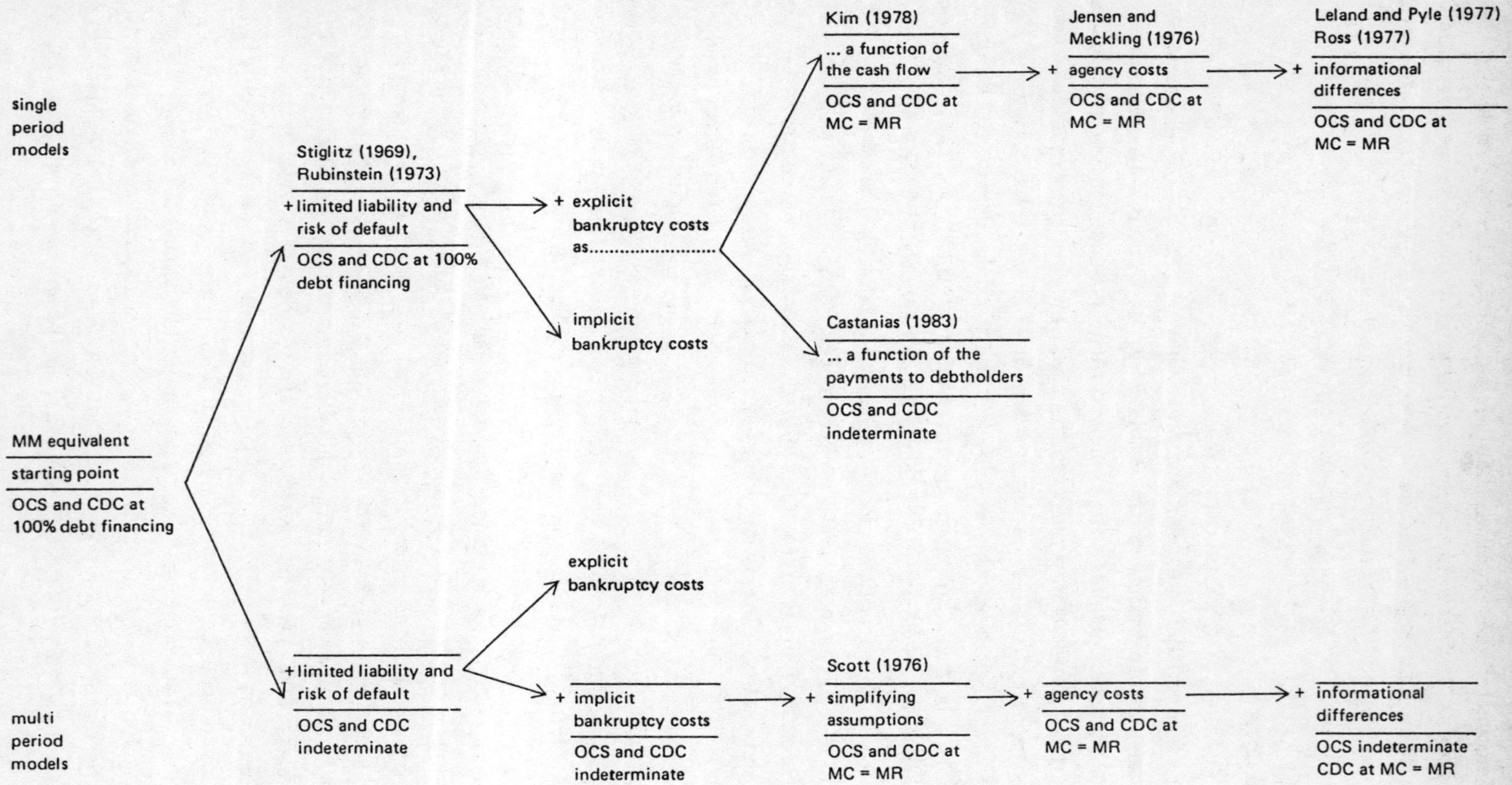


note table 4.1 OCS = optimal capital structure
CDC = corporate debt capacity
MC = marginal costs
MR = marginal revenue

determinate results (although not in a "closed form"). However, if a slightly different bankruptcy condition would be used in the Scott–model, i.e. if the tax rebate would be omitted from it, the results would become indeterminate.

Hence, it can be concluded that the robustness of the various theoretical models leaves much to be desired, so that the prospects of extending the models with more realistic assumptions are questionable, to say the least. The complexity and indeterminate results of the models are presented with some emphasis, because these aspects are hardly reported on in the literature. Virtually all published models are "well behaved".

However, the rather unsatisfactory conclusions above do not alter the fact that a number of models with determinate results is obtained, in which various aspects of the debt financing decision are incorporated. In these models, the firm's optimal leverage (optimal capital structure) and maximal leverage (corporate debt capacity) is seen to be:

- positively influenced by the tax rate;
- negatively influenced by the bankruptcy costs (implicitly or explicitly specified) and the agency costs of debt;
- ambiguously influenced by the risk free rate of return and the characteristics of the distribution of the cash flow.

These theoretically derived influences can be fruitfully used as hypotheses in empirical analyses.

As a final remark it is noted, that the value of the theoretical exercises should not only be measured by their direct contribution to the empirical analyses. They have an indirect effect as well. Analysing firm finance at such a high level of abstraction focuses the attention on the essential elements of the leverage decision and describes the problem in terms of broad categories of debt ratio determinants. This provides us, so to speak, with a vantage point from which the many empirical variables and suggestions from the every day practice of business finance can be seen in their proper perspective. This effect can pervade all empirical analyses and, thus, be more significant than the direct contribution in the form of hypotheses.

Table 4.1A List of symbols

| symbol | content |
|---|---|
| $Y_e$ | end of period value of equity |
| $Y_d$ | end of period value of debt |
| $\tilde{x}$ | cash flow before interest and taxes |
| $f(\tilde{x})$ | probability density function of $\tilde{x}$, assumed to be continuous and twice differentiable |
| $r$ | risk free rate of interest |
| $\tau$ | corporate tax rate |
| $V_e$ | value of equity |
| $V_d$ | value of debt |
| $V=V_e+V_d$ | firm value |
| $b$ | bankruptcy condition |
| $B(.)$ | bankruptcy costs as a function of (.), but no greater than (.); assumed twice differentiable with positive first derivative, second $\geq 0$; always positive in the bankruptcy zone, zero otherwise |
| $F$ | probability of bankruptcy, i.e.: $\int_{-\infty}^{b} f(\tilde{x})d\tilde{x}$ |
| $R$ | payment to debtholders |
| $A(.)$ | agency costs as a function of, but no greater than (.); assumed twice differentiable with first and second derivative $\geq 0$ |
| $E(.)$ | expectation operator |
| $\tilde{.}$ | random variable |
| $\bar{.}$ | expectation/average |
| $L$ | liquidation value of the firm's assets in imperfect secondary markets ($L<V$) |

# 5 DETERMINANTS FROM THE PRACTICE OF SMALL BUSINESS FINANCE

## 5.1 Introduction and overview

Because of its nature, the practical literature is much less easy to inventory than the theoretical literature is. In the latter, determinants are derived in a well defined framework of assumptions, objectives and instruments, aimed at reducing phenomena to their absolute essentials. The result is a number of neatly arranged categories of determinants. The practical literature, however, lacks the guidance of such a strict framework and the result is a myriad of different aspects of the firm's financial behaviour. The subjects addressed in this literature vary from the start of a small business to the prediction of corporate bankruptcy and the methods used to tackle these problems include empirical facts, rules of thumb, practical guidelines and research outcomes as simple correlations and the results of multivariate techniques. The practical literature also covers a wider area and includes not only all debt and equity categories but also the different types of assets. The fragmented nature of this literature will almost inevitably be reflected in the discussion below, but fortunately, some useful recapitulations of the practical literature have been published which help to map this vast area. In particular, the studies by Sprenger et al. (1982), Van der Linden et al. (1988) and by Walker and Petty (1978a) are used to draw up the summary in the following subsections. The stratification of the influences on financial structure used in the following is adopted from the study by Sprenger et al. (1982), in which the following categories are distinguished:

1. influences pertaining to the individual firm:
   –characteristics of the firm itself;
   –characteristics of the entrepreneur;
2. influences pertaining to the relationships among firms;
3. influences pertaining to trade and industry in general.

In order to give some overview, a number of determinants of financial structure suggested in the practical literature is schematically depicted in table 5.1. It should be noted, however, that the selection in table 5.1 is rather arbitrary and other items could be mentioned as well.

Table 5.1 Overview of financial structure determinants in the practical literature

Financial structure determinants pertaining to:

relationship among firms
- banks
- customers
- suppliers
- competitors

economy in general
- state of trade
- government:
  - legislation
  - taxes and subsidies
- financing rules

individual firm

firm itself
- size
- legal form
- industry classification
- financial situation
- collateral security
- life cycle

entrepreneur
- degree of information
- age/succession
- attitudes towards:
  - risk
  - debt
- objectives:
  - profit
  - independence

In the following two subsections these influences, which are collected from a number of sources, are summarized within the above framework. Beforehand, however, some introductory remarks seem to be in place. First of all, for the sake of brevity, parsimony is practiced in making references to the literature. Ample references can be found in the three studies mentioned above, and the broad variety of little facts, that is incorporated in the sequel, makes it almost impossible to place a reference at each fact. Secondly, the determinants of financial structure suggested in the practical literature are, of course, neither completely disjunct from the ones developed in the theory, nor totally unrelated to the characteristics of the small firm's capital markets. Hence, a number of determinants already received some attention in the foregoing, particularly in section 2.4 and in the chapters 3 and 4. This allows the discussion in the present chapter to be kept short. Thirdly, rather than sifting out the crucial determinants, this summary presents the influences on financial structure as they are put forward in the literature, ripe and unripe mixed together. This procedure allows the multitude of arguments to come to the fore, although it will probably also express some of the lack of coherence in this literature. In the next chapter, an attempt is made to compare and evaluate the findings from the different sources.

## 5.2 Determinants related to the firm's internal characteristics

Characteristics of the firm

For obvious reasons, the characteristics of the firm, particularly its financial and operating characteristics, have always been an important basis for decisions regarding its financing. Indeed, many of these characteristics have a direct relation with the outcome of financial contracts so that, consequently, in the practice of business finance there is an urgent want of procedures to arrange and combine the various characteristics into a concise set of relevant figures. In response to this want, an enormous quantity of literature has been brought out in this area of inquiry, which is generally denoted as financial statement analysis. The main topics in financial statement analysis are repeated over and over again in textbooks and they are an obligatory subject in almost every business course. So it is not very surprising, that the similarities in this literature are large[1]. Firm characteristics are usually arranged in the categories liquidity, leverage, profitability and activity, while the methods used to combine them almost invariably involve ratio analysis (or one of its variants, such as "common-size statements" or "trend statements"; see Foster, 1978). Aspects of financial statement analysis are discussed later on in this study. In the present chapter, attention centres on the determinants of financial structure that can be collected from this literature. This is a somewhat peripheral aspect of financial statement analysis, which is more concerned with the "internal relations" in financial statements than with outside determinants. Nevertheless, the following determinants can be mentioned:

Firm size is suggested as a determinant of financial structure in almost every study and for a number of different reasons. First of all, firm size may determine the degree to which a firm can call on the capital market (cf. section 2.4). Small firms may lack access to organized capital markets,

-------

[1] *According to Horrigan* (1968), *the bulk of this literature consists of instructions on how to compute ratios.*

both for bonds and stock. If they have access, the flotation costs of selling a small issue may become almost prohibitive. Fixed costs in the $30,000 to $50,000 range are reported on; other estimates are that these costs may approach 25% of the issue size. If an issue has taken place, the small firm's securities may lack marketability which makes it necessary to price the securities so that they will yield a significantly higher return than those of a similar firm having a ready market for its securities. In the practical literature, the small firm's access to the capital market is described as an arduous hurdle-race indeed. Secondly, firm size may determine the bargaining power in financial contracting. Large firms usually can choose from a wider variety of debt forms, including favourable special offers, than small firms. The larger amounts involved make it worth while to tailor the debt contract to the preferences of both parties, instead of using standard debt forms. Thirdly, there may be scale effects in costs and returns which enable larger firms to retain more earnings. Finally, the following size related characteristics are also reported to influence financial structure: small firms frequently have no specialized staff, do not use financial planning and have not developed their accounting system into a management information system.

Legal form also determines the access to capital markets, since only corporate bodies can be listed. Additionally, legal form may determine the possibilities of outside participations, such as limited partnerships. Moreover, depending on the legal form, various conflicts of interest between stakeholders in the firm may arise, affecting its financing. Active participants may prefer to retain earnings while non-active participants may prefer dividends to be paid. Management may want to issue new common stock which can cost the existing shareholders a portion of their net worth if the price of the new issue is not carefully established. Also, issuing common stock to external investors permits a potential dilution of control for existing shareholders. Although such a loss may be only psychological (covenants in a debt contract can have a similar effect), experience indicates that most small business owners are quite averse to placing them in a position that even slightly compromises their control over the firm. Finally, since dividends are not mandatory, corporate bodies may pass the dividends in a bad year without experiencing legal failure, while the owner-manager must use some returns for his cost of living. Thus, legal form can provide increased latitude.

Industry classification is a possible determinant of financial structure on several grounds. It is pointed out (among other things in a brochure of a commercial bank) that industrial firms are comparatively capital intensive because they require large investments in fixed assets[2]. This negatively affects their flexibility and, hence, they require a comparatively large proportion of equity. Industry classification can also be related to volatility in both earnings and asset requirements which, in turn, influences financial structure. As cyclical variations in asset requirements are encountered, short term debt offers the advantage of flexibility. This permits the firm to keep funds fully invested during slack periods by reducing company liabilities. In more stable circumstances, long term debt may be preferred because it avoids frequent renegotiations of loans; equity avoids the refinancing process altogether since a maturity date is not forthcoming[3]. A direct influence of industry classification occurs when lenders check the state of the market in the industry before granting a loan. This would allow firms in growing industries to obtain credit more easily than firms in contracting industries. Perhaps because of all the above reasons taken together, there seem to be, for a long time past, different "target-ratios" for different industries, that are applied without further motivation.

The financial situation of the firm is also mentioned as a determinant of financial structure, although this argument is very near becoming circular. More specifically, it is stated that a low proportion of equity is associated with more difficulties in obtaining additional debt, equity and outside participations, with the use of more trade credit, with the use of more loans from friends and relatives, etc. An additional, and more relevant, aspect of the firm's financial situation is the effect on the debt ratio of recent investments that required a large single sum. It is reported that, in small business, these investments usually cannot be financed with the current proportions of debt and equity in a firm. The reason for this is, that retained earnings are the main source of equity for small firms and the funds that can be internally generated, at least within a short period, usually are not substantial enough. Hence, recent substantial investments are associated with temporarily high debt ratios. Conversely, if substantial investments are planned for the near future, debt ratios may be kept low by retaining earnings in advance, in order to accumulate reserve borrowing power.

---

[2] *The scant empirical evidence does not support this contention unequivocally (see Van Iperenburg and Van der Wijst*, 1985)

[3] *Some authors, e.g. Walker and Petty* (1978*a*), *hold a different view and state that short term financing should be curtailed in highly competitive or volatile industries (see table* 5.2).

The collateral security is generally considered to be an essential determinant of the possibilities to use debt financing. "Tangible assets enable a firm to borrow", is the title of one of the paragraphs in the 1987 report of the president on the state of small business in the USA. According to some authors, collateral makes debt financing easier because the lender can look to the personal assets of the owners as well as to the cash flow of the company. Small firms are frequently considered to be limited in their opportunities to use debt, especially long term debt, because of an insufficient amount of available collaterals. In this context it may be noticed, that in certain parts of the retail trade the inventories can serve as collateral for loans (e.g. in supermarkets) while in other parts banks do not accept inventories as collateral (e.g. ladies' and men's wear).

Life cycle effects are frequently mentioned in the practical literature. Usually, the firm's life cycle is depicted as an S–shaped (normal) curve, with a starting small business at the beginning, a large, listed firm at the top of the curve and bankruptcy or take–over at the end. This curve is divided into several "life–stages" that, allegedly, each have their typical financial problems. Most life cycle studies have in common that they do not include supporting empirical evidence, while almost each author uses his own division in "typical" life stages[4]. However, life cycle effects usually boil down to the following. At the very start of a business, personal savings and friends and relatives are the primary source of capital. As size (and age) grow, various types of debt (first short term, then long term) gradually become more important, until the firm is large enough to enter the public issue market and obtain outside equity (see Van der Linden et al., 1988, for a more comprehensive discussion). The latter effect is of course not very relevant for small business.

A different life cycle effect is described by Mueller (1972), based on the argument that, given the high income and wealth positions of top managers, a manager's primary objective will be to expand or maintain the size of the company, rather than to maximize stockholder welfare. This would induce managers to continue investing, even if stockholder welfare would require to disinvest. Mueller argues that this effect is primarily relevant for large, mature firms, in which a substantial internal flow of funds is at the management's discretion. An empirical implication of this argument would be, that these firms would prefer to use internally generated funds for their

-------

[4] *Compare, for instance, the five stages of small business growth discussed by Churchill and Lewis* (1983) *with the three primary financing stages that are described by Walker and Petty* (1978*a*).

investments rather than debt or outside equity, which is an interesting parallel to Myers' pecking order effect.

In the foregoing, the influence of firm characteristics on financial structure is primarily discussed in terms of leverage. However, the maturity structure of debt can also be influenced by firm characteristics. A summary of these influences, some of which are already mentioned, is presented in table 5.2.

Table 5.2 Factors influencing the choice between short and long term debt

| Firm characteristics affecting the maturity structure of debt | Factors which favour selecting | |
|---|---|---|
| | long term debt | short term debt |
| Structure of the industry | Cyclical or highly competitive | Noncyclical or moderate competitive |
| Company degree of operating leverage | high | low |
| Existing asset structure | high fixed asset component | high current asset component |
| Company size and maturity | large and well established | small and newly established |

*Source: Walker and Petty* (1978*a*), *adapted.*

## Characteristics of the entrepreneur

In the majority of small firms, less than a dozen people is employed and many small firms have no employees at all. With such a small work force, the entrepreneur not only bears the ultimate responsibility, but he also accounts for a large number of other tasks. This dominant position of the entrepreneur in the small firm induced many researchers to study his personal characteristics as an explanatory factor for firm success. The first studies focussed on the "entrepreneurial characteristics", i.e. traits of character which are rooted in the social background and early experiences of the person (e.g. failure at school, high need for achievement etc.). However, this approach appeared to produce "stereotype" descriptions of entrepreneurs, which are only relevant in extreme cases. Hence, attention gradually shifted from "entrepreneurial characteristics" to "entrepreneurial functions" and "entrepreneurial performance", while more and more elements from the socio-economic environment were included in studies of entrepreneurship (see Van der Linden et al., 1988, for a more detailed discussion). In spite of these research efforts, however, the successful entrepreneur is still not easy to describe. This is illustrated by Kilby who, in his article "Hunting the heffalump", compares the successful entrepreneur with the monster of Loch Ness: most people think it exists, but nobody can give an accurate description (Kilby, 1971, quoted by Van der Linden et al., 1988).

From the above it will be clear, that the characteristics of the entrepreneur are not likely to constitute a solid basis for analysing differences in financial structure. However, a number of separate correlations is reported on in the literature and these are briefly summarized below.

The degree to which the entrepreneur is informed is likely to be of importance for every decision within the firm. As regards finance, empirical research indicates that entrepreneurs, who are better informed in terms of general education and knowledge of leasing and leverage, use debt more often than their less informed colleagues. For instance, artisans are reported to neglect the possibilities of debt financing owing to a lack of commercial knowledge. Moreover, in some publications the quality of financial management in SME's, particularly the capacity to overcome financial difficulties, is partly attributed to the degree in which the entrepreneur is informed.

The age of the entrepreneur may influence the firm's financial structure, along with the arrangements made for his succession. From a certain age up, entrepreneurs are no longer willing to invest in their firm. Also, older entrepreneurs use less debt and are less willing to accept outside participations than young ones. An exception are the older entrepreneurs without successor; they are willing to accept outside participants as possible successors.

The attitudes towards business risk in general and debt financing in particular are also suggested as possible determinants of financial structure. Research outcomes indicate that entrepreneurs who are highly risk averse use comparatively little debt, while a positive attitude towards debt financing is associated with a high degree of leverage. Similarly, dynamic entrepreneurs are reported to invest more than their less dynamic colleagues. A host of other specific correlations can be found in the literature, but in essence they boil down to the above arguments.

The entrepreneur's objectives are reported to relate to the firm's financial structure in several ways. Entrepreneurs are likely to include various aspects of financing instruments in their decision making process and to evaluate each instrument in the light of their preferences. For instance, some empirical studies indicate that entrepreneurs with a strong preference for independence may reject ways of financing that involve a regular supply of information to parties outside the firm (e.g. outside participation). Similarly, the objective of independence seems to make small firms use more equity (as long as it does not involve outside participation) and retained earnings than the objective of profit does. Conversely, firms that seek profits or growth use a comparatively large proportion of debt.

## 5.3 Determinants related to the firm's external relationships

Banks and financial institutions may contribute to financial difficulties in small firms by negligence in the loan negotiation process, e.g. by insufficient solvency checking and by overvaluing management potential. A German study attributes a part of the financial difficulties to banks and financial institutions in one out of every three cases of insolvency (see Sprenger et al., 1982, p. 44). On the other hand, banks and financial institutions may influence the financial policies of small and medium-sized firms by imposing too strict solvency requirements. This may especially be the case in times of a tight money policy by the central bank, as some evidence indicates. The third possibility, i.e. that the bank's requirements are neither too strict nor too liberal, is apparently so hypothetical that it is not mentioned in the literature.

Suppliers, customers and competitors co-determine the conditions on which business is done and this may affect the financial situation of SME's in two ways. On the one hand, suppliers may influence the financial structure of their customers by the terms on which they grant trade credit. A broad variety of terms is encountered in practice, ranging from suppliers who also provide long term debt (in order to support the sale of e.g. expensive machinery) to suppliers who insist on cash payments. The interest to be paid after the period of trade credit also varies across suppliers. On the other hand suppliers may increase total investments by imposing a large minimal order quantity and a long time of delivery. The same arguments, but in the opposite direction, apply to the influence of customers on the financial structure of small businesses. The intensity of the competition in the buying and selling market determines a firm's bargaining power and, hence, the extent to which it can profit –or must suffer– from the prevailing terms of trade credit. Additionally, fierce competition may erode profits which negatively influences the possibilities to retain earnings.

The state of trade (phase of the business cycle) is reported to have an impact on financial structure for several reasons. First of all, firms and consumers act differently in different phases of the business cycle: during a recession, customers are less able to pay, hence, accounts receivable increase and a larger proportion of the accounts becomes dubious. Secondly, decreasing demand during a recession tends to increase inventories if production cannot be reduced. If production can be reduced, the danger of idling expensive machinery arises, which decreases profits and, hence, diminishes the possibilities to retain earnings. Thirdly, for banks and financial institutions money is comparatively tight during the revival and boom phase of the business cycle, and money is comparatively easy during the recession. Equilibrium will be maintained by varying the interest and non-interest conditions on which loans are granted, i.e. the debt policy of banks and financial institutions depends on the phase of the business cycle. Fourthly, the number of bankruptcies depends on the business cycle, which makes investing and financing more risky during the recession phase. The increased risk may have effects that extend well into the revival phase. There may be a considerable psychological effect on decision makers, who become much more cautious if they experienced a large number of bankruptcies among their business connections. Also, the formal procedures and conditions in e.g. banks may be sharpened considerably in reaction to increasing bankruptcies, while these regulations are not automatically relaxed as the state of trade improves.

Government measures and regulations affect virtually every aspect of a firm's operations and most, if not all, of these measures are likely to have a direct or indirect impact on the firm's financial structure. However, in the literature little or no attempt is made to assess the financial consequences of the indirect influences and only the most obvious direct effects are discussed. Taxes are an obvious government influence and the fiscal legislation in force in most Western countries taxes profits and considers interest payments a tax deductible item. This diminishes the possibilities to retain earnings and benefits the use of debt. Besides the monetary effect, fiscal legislation also imposes a considerable administrative burden upon firms. This effect is also inherent to the extensive system of social securities called into existence in some European countries. Much clerical work is, however, not an exclusive characteristic of government levy; government subsidies on certain costs, investments and loans also are associated with a lot of red-tape. Moreover, the sheer number of different subsidies makes expert advice virtually indispensable for small firms. For instance, in The Netherlands government influence on the business community is blossoming into no less than a hundred different subsidies with no central office to apply for them. The discussion in the literature of government influences is somewhat scant, perhaps because of the multitude of different measures that usually apply equally to all firms, while each individual measure has only a minor effect on financial structure.

Rules of finance are also an element of the firm's environment and their influence could well be larger than it appears to be at first sight. An obvious influence occurs when banks have internal rules which stipulate that credit should only be supplied up to a certain degree of leverage, but such rules are, of course, hidden from view. More perceptible are the traditional rules of finance that, over many decades, have been proclaimed in textbooks and have been taught at business schools and courses. Perhaps the most famous of these is the rule that the firm's current assets should be twice its current liabilities. This two to one current ratio rule probably had a more significant and longlasting impact on financing practices than any other rule (Horrigan, 1968). Similarly strict rules, that once were in use, vanished more quickly. For instance, in the early twenties Polak discussed rules from German origin which stipulate that the firm's inventories should not exceed its long term debt and equity, whereas the trade credit obtained from suppliers should not exceed the credit granted to customers (Polak, 1950, p. 179). Today, few people would comply with (or even remember) these rules. Other rules are more persistent though. The so-called "golden balance sheet rule", which prescribes the synchronization of the terms of investing and financing, is still in use today, as appears from table 5.2. Moreover, research by Donckels et al. (1987) reveals that this rule is the stated financial policy of three fourths of the Flemish SME's. Thus, traditional rules of finance should be kept in mind when analysing small firm financial structure.

## 6.1 Comparison and evaluation

Looking back on the discussion in the previous three chapters, the first impression that is likely to arise is that the determinants of financial structure presented in the theory of finance are rather different from the ones suggested in the practical literature. Indeed, at first glance the plethora of different subjects and methodologies incorporated in the practical literature seems to bear little resemblance to the neatly arranged theory based determinants. However, on closer investigation certain common elements become clear. In some cases, such as the effect of collateral security on the financing decision, the theoretical and practical literature obviously describe the same determinants of capital structure. In other cases, the practical literature refers to phenomena that are usually assumed not to exist in financial theory, e.g. transaction costs and limited access to capital markets. In still other cases, such as the entrepreneur's age, the relation between the two is less obvious but can still be argued to exist (viz. by the nondiversification argument which may lead to "age related" financial policies). Additionally, the growing attention in the theory of finance for empirical aspects of the firm's capital structure has induced theorists to draw upon the empirical data previously dominated by pragmatists. If anything, both approaches share the same empirical basis.

Since these common elements mainly refer to the phenomena being studied, a first conclusion can be that both approaches differ more in method than in content. There are marked differences in the methods applied. As could be expected, the theoretical approach rests on the application of strict rules to derive conclusions from the assumptions, whereas the practical literature arrives at conclusions using many different methods, including reasoning, empirical facts and research outcomes such as simple correlations. As a consequence, the nature of the conclusions differs rather strongly between both approaches. In the theoretical approach, simple and straightforward conclusions are derived on the basis of a large set of complex assumptions. By contrast, in the practical literature complex and often entangled conclusions are formulated on the basis of simple, if any, explicit assumptions[1].

In view of the variety of methods and the nature of the conclusions in the practical literature, it is perhaps not surprising to detect a number of flaws in the various argumentations. In the previous chapter's summary, there are several cases in which determinants of financial structure, once reviewed within a "cause and effect" framework, appear to be motivated in a somewhat questionable manner. First, causes are put forward without a corresponding effect, e.g. small firms are reported to face high flotation costs and to lack a specialized staff, but the effects on the firm's leverage remain unclear. Second, causes are presented that fail to capture the purport of the argument in practice. For instance, legal form is argued to determine the possibility of outside participations. Although legally this is a correct argument, it is much more likely that in practice legal form is chosen to accommodate the available sources of finance (i.e. cause and effect are reversed). Similarly, corporate bodies may indeed pass the dividends in a bad year, but this only provides increased latitude if the owner-manager has the disposal of other sources of income. If the owner-manager depends on the firm for his livelihood, he must use some of the earnings for his cost of living, regardless the legal form. Third, in the practical literature simple correlations are encountered without an accompanying cause and effect motivation. For instance, younger and better informed entrepreneurs are reported to use more debt solely on the basis of statistical correlation. Fourth, and last, we find in the practical literature examples of two or more causes mixed together. For instance, small firms allegedly use little debt because of insufficient collateral and low profitability. In this argument, the effects of size, collateral and profitability are mixed. The above discussion of various flawed argumentations is not intended to disparage the practical literature, but rather to accentuate the need of a critical evaluation of its conclusions.

One aspect of the practical literature that is often overlooked, is that the concepts and techniques used in these studies are usually quite old, some even date back to the beginning of the century. For instance, the classical example of a ratio pyramid, the DuPont system of financial analysis or DuPont chart, which is still widely used today, was developed in 1919 by the American company of the same name. Early studies of "credit barometrics" and of financial differences between successful and unsuccessful firms date

-------

[1] *The assumptions made implicitly in the formulation of conclusions can, of course, be quite complex.*

from the same period, although the introduction of new statistical techniques, such as multiple discriminant analysis, brought about that they were partly reinvented in the sixties[2]. In view of the long history and the large number of studies, a surprising aspect of this literature is, that a stable body of knowledge never materialized. There is still no agreement on e.g. what variables are good "barometers" or what the target values in ratio pyramids as the DuPont chart should be. In this respect, the "practical value" of the practical literature is limited, i.e. different studies are likely to give different answers to these questions (cf. the variety of scale effects reported in table 2.3).

Such contradictions are not encountered in the theoretical literature which, however, has its own limitations. As the analyses in chapter 4 show, most theoretical models are rather sensitive for small changes in their specification. Particularly in the multi-period context the range of specifications that produce usable models is discomfortingly narrow. This indicates that much theoretical work still needs to be done, even within the comparatively simple and well-known framework of the extended MM-theory. From the analyses this theory appears to be somewhat less solid than is generally suggested in the literature, perhaps because publications tend to have a strong bias for positive research outcomes.

However, well-behaved models are also obtained and their comparative statics provide a number of hypotheses for the empirical analyses. These hypotheses are formulated in qualitative terms, i.e. they specify the direction of an influence which, subsequently, can be tested with statistical methods, along with other hypotheses that formulated on different grounds. In this procedure, the theoretical exercises are, in qualitative terms, connected with the empirical explanation of observed capital structures. In view of the complexity of human decision making in a dynamic and uncertain financial world and the practical impossibility to incorporate more than a few aspects of this complex reality in any model, this procedure is in our opinion quite satisfactory. We consider the contribution of the theory to this study, although it is limited to qualitative aspects, to be a valuable one.

This opinion can hardly be regarded as a consensus view within the profession; others arrive at more critical or more optimistic conclusions. For instance, in his 1983 presidential address to the AFA Myers (1984)

-------

[2]*See Horrigan* (1968) *or Van der Wijst* (1985) *for a more detailed overview of these developments.*

answers the question "How do firms choose their capital structures?" by stating: "We don't know.". Similarly, in a critical and well documented treatise Tempelaar (1985) discusses the degree in which valuation formulae, such as those developed in chapter 4, are operational. In his opinion this degree is virtually nil, since the formulae only lead to rough and qualitative conclusions; there is no question of determining the theoretical value of corporate claims in quantitative terms. Tempelaar is disappointed by these results of some thirty years of research and he wonders what the results would have been had the same research efforts been applied in a different direction. Like many authors, he refers to the work of Donaldson (1961, 1969) as a possible alternative for the neoclassical theory of finance. On the other hand, Jonkhart (1980, p.112) shows a more optimistic point of view as he states that there is a pretty good chance that the MM benefit–cost model of the capital structure, if properly extended to include all relevant capital market imperfections, ultimately may provide a full explanation of observed leverage ratios.

In our opinion, different aspects of these points of view are relevant for this study. On the one hand, the value of the theoretical approach for the analysis of financial structure in small business is not so much that it gives a quantitative explanation of the actual financing patterns, as it ultimately might do for large, listed firms, but that it provides an analysis of the firm's financial behaviour on a high level of abstraction. This gives an insight into the general lines of financial policy, albeit in qualitative terms, and prevents the analysis from petering out in the thousand little details inevitably associated with financial contracting. The practical literature, on the other hand, occupies itself with the details of business finance on a low level of abstraction. This provides an insight into the possible effects of phenomena not incorporated in the theory and prevents the analysis from disregarding potentially important determinants of observed financial structures. Although this is also a valuable contribution to this study, it should be remembered that seventy years of research in this direction failed to produce a stable body of knowledge. Without the guidance of a theoretical framework it appears to be difficult to draw inferences from the multitude of research outcomes[3]. Thus, as an overall conclusion, the theoretical and practical approach can be inferred to supplement each other, both in describing the financial behaviour of (small) firms and in providing a basis for empirical analyses.

-------

[3] *In this respect, the most practical thing to have would be a good theory.*

## 6.2 Summary and empirical implications

The previous section's conclusion, that the theoretical and practical approach largely describe the same phenomena but from a different point of view, can be illustrated by combining the various determinants into a single summarizing overview. Table 6.1 gives such a summary, which shows the similarity in phenomena studied. Together, the theoretical and practical literature provide a considerable number of hypotheses, which are summarized in the columns "empirical (proxy) variable" and "expected effects" of table 6.1.

As regards the application in empirical analyses, these determinants have in common that they cannot be used without some precaution, albeit for different reasons. It is in the nature of the practical literature that its conclusions are frequently drawn up in the context of a specific research project or for a specific group of firms. For lack of underlying theory, the validity of the conclusions is limited to the same specific areas and generalizations should not be made without precaution. On the other hand, the theoretical literature only produces general results, which, however, cannot be applied in a specific situation without the precaution of checking the aptitude of the underlying set of assumptions. Phenomena of importance in a specific situation may be "assumed away" in theoretical analyses. This is particularly true for analyses in small business as most of the theoretical literature implicitly or explicitly alludes to large listed firms. With these precautions, however, the determinants in table 6.1 may constitute a solid basis for the empirical analyses, which are the subject of the following two chapters.

Table 6.1 Summary of financial structure determinants

| determinants from practice | corresponding theoretical determinants | empirical (proxy)variables | expected effects |
|---|---|---|---|
| INDIVIDUAL FIRM | | | |
| - firm size | size | earnings, assets, sales | positive on debt |
| - legal form | complexity of ownership structure | >1 corporate body limited liability | negative on debt |
| | co-ownership | >1 owner/manager | positive on debt |
| - industry classification | durable effect (Titman) | ind. classification dummy | negative on debt |
| - financial situation of the firm and/or the industry w.r.t.: | | | |
| o growth | growth (Myers' effect) | change in total assets, sales (R & D exp.) | positive on short term debt<br>negative on long term and total debt |
| | growth (Signal effect) | change in total assets, sales (R & D exp.) | positive on debt |
| o profitability | profitability (pecking order effect) | r.o.i., net profits,accumulated retained earnings | negative on debt |
| o volatility of earnings | volatility of earnings | statistical dispersion of earnings | negative on debt |
| o volatility of asset requirements | | cyclical (seasonal) variation in assets, sales | positive on short term debt |
| o recent substantial investments | - | amount of recent investments (fixed assets) | positive on (long term) debt |
| - collateral security | liquidation, collateral value | asset structure, inventory turnover | positive on debt |
| | private collateral, fraction of the owner's wealth held in the firm | private collateral, industry classification dummy (inventories accepted as collateral) | positive on debt |
| - life cycle | - | age of the firm | positive on debt |
| ENTREPRENEUR | | | |
| - age | non-diversification | age of the entrepreneur | negative on debt |
| - degree of information | - | level of education | positive on debt |
| - attitudes/objectives | - | - | - |
| AMONG FIRMS | - | - | - |
| IN GENERAL | - | - | - |
| - government | tax advantage of debt | depreciation, tax credits (R&D, advertising) | negative on debt |

# PART III
## empirical analyses in small business

# 7 ANALYSES OF SAMPLES OF INDIVIDUAL FIRMS

## 7.1 Introduction

The third part of this study is concerned with two series of empirical investigations, in which data on different levels of aggregation are analysed. Samples of individual firms within more or less narrowly defined industries are analysed in the present chapter, as are two databases of industry averages in the next. Testing the expected effects of the empirical variables associated with the determinants of financial structure in table 6.1 may seem to be a relatively straightforward matter, but some caveats are in place. Empirical tests of hypotheses, which are derived from theories as those outlined in the previous chapters, are hampered by a number of discrepancies between theory and practice. Theories are built on assumptions or hypothetical constructs, which may not occur in practice or only to a limited degree. For instance, the empirical implications of theories of optimal capital structure are based on hypothetical constructs regarding the functioning of the capital market. Moreover, theoretical concepts may have no practical equivalents or the equivalents may not be available in the data. Theoretical valuation formulae e.g., are derived using expected future cash flows which are impossible to measure. Market values are measurable, but in many studies, including this one, book values have to be used instead. Hence, when we test a hypothesis, we usually not only test an empirical implication of the theory, but also constructs on which the theory is based and even the adequacy of the data. In testing hypotheses we pose several questions at the same time, so to speak. For instance, it is well known from Roll's critique that the empirical validity of the CAPM cannot be tested without simultaneously testing for the efficiency of the market portfolio.

These and similar problems contribute to the complexity of empirical tests and make the results difficult to interpret, since failure to establish an expected result may have several causes. This is even aggravated by the fact that the "proper" data are usually not available, so that approximations have to be used. The latter may evoke interpretations, in which expected results tend to be attributed to theoretical considerations, while unexpected results are blamed on data problems. In order to avoid such interpretations, the estimation results in this study are parsimoniously commented upon, particularly regarding the data. As is mentioned before, the data used in this study only comprise book values, which can differ from the corresponding market values. However, book values and market values are, on average, likely to be strongly correlated, particularly as regards the "current" items on the balance sheet. In the industries investigated in this study, inventories, accounts receivable and payable, other short term debt, etc. all have a high speed of turnaround. This means that these items are continuously renewed at market prices, so that the difference between book value and market value will be small. For the fixed items the difference may be larger, but on the whole we feel that the data provide a firm footing for empirical analyses and applications.

In the next section, we will first describe the samples of individual firms that are analysed in this chapter. Subsequently, we focus in section 7.3 on the hypotheses and variables, followed in section 7.4 by the specification and estimation results.

## 7.2 Data

The data analysed in this chapter refer to individual firms within more or less narrowly defined industries; the unit of observation used here is the individual firm. Compared with the industry averages used in the next chapter, individual data are likely to be characterized by a larger degree of dispersion. The effects of e.g. cumulative depreciation and the personal wealth of the entrepreneur, which can safely be assumed to average out at the industry level, will be reflected in individual data. Also, certain variables that can be fruitfully used with averaged data may have little relevance here, because they vary strongly between but usually considerably less within shop types (e.g. the nature of the principal items in the assortment). On the other hand, investigating samples of individual firms permits the data collection to be tailored to the specific characteristics of the industry, so that a more accurate description can be obtained. In most of the samples analysed in this chapter, there is a much wider variety of variables to choose from than is the case in the analyses of industry averages.

The data used in this chapter come from two sources. Most samples (12 out of 14) are collected by the EIM, the Research Institute for Small and Medium-sized Business in the Netherlands. With the exception of structural engineering, all samples refer to the retail trade. These date are collected for several reasons, two of which are relevant for this study. First, one of the purposes of the data collected efforts is to provide a detailed and accurate basis for interfirm comparisons. This means that financial statements, as well as many other variables, are included in the data of each sample. It also means that the very small and very large firms are excluded from the samples because they are considered (by the EIM) to be outlyers in the context of interfirm comparison. Second, the data are collected to provide information on one or more problems of current interest within the industry[1], which means that no two samples are identical in terms of variable composition. As a result, the number of financial and operating characteristics that can be used for the explanation of financial structure differs between the samples, depending on the nature of the problems investigated[2].

All samples of the EIM have in common that the data are collected by the institute's own field force of trained analysts. This, of course, contributes substantially to the value of the data. For instance, the uniformity of the various items in the financial statements leaves nothing to be desired, while private elements are carefully removed from the business accounts. Further, the valuation of certain items on the balance sheet (particularly buildings and inventories) is adjusted if the original valuation is kept low in order to evade taxes. The trained analysts also provided accurate estimates of data that usually are not recorded, such as shop surface. Such painstaking collection methods cannot fail to produce high quality data sets. Regrettably, in 1986 the institute stopped these data collection activities and abandoned its field force.

The remaining two samples, referring to the metal products industry and the printing industry, are made available by Rabobank Nederland, one of the largest banks in the Netherlands. Like most other banks, Rabobank Nederland more or less regularly collects data that, on the one hand, are used in the loan evaluation process by local branches and that, on the other hand, are published in "industry studies" as a service to the public at large[3]. Because banks have clients in all sectors of the business community, they cannot define "industries" as narrowly as the EIM does, since that would lead to a very large number of different industries. Also, the information collected for each firm is confined to the financial statement and some

-------

[1]*The nature of these problems varies strongly; they include legal, commercial, social and financial problems.*

[2]*All samples used here are reported on in separate "industry studies" which are published by the research Institute for Small and Medium-sized Business in Zoetermeer, The Netherlands.*

[3]*See, for instance, "Bedrijfstakstudie Grafische Industrie" published by Rabobank Nederland, Stafgroep Economisch Onderzoek, Bedrijfstakonderzoek, Utrecht* 1984.

supplementary data regarding e.g. legal form, labour volume etc. In spite of their somewhat limited information per firm, however, the two samples provided by Rabobank Nederland are a valuable contribution to this study because, unlike most of the EIM data, they refer to non-retailing industries and to comparatively large SME's.

Some key figures of the samples used in this chapter are summarized in table 7.1. The sample sizes in this table refer to the number of firms actually used in the analyses. The original samples are (sometimes considerably) larger because branches of chain stores are included. Since these branches are not independent financial units (they have no balance sheets), they are excluded here. Total assets, as all other items of the balance sheet used in this chapter, refers to the average of beginning and ending balance sheet. The other figures in table 7.1 largely speak for themselves.

Table 7.1 Some key figures of the samples

| Sample | year | sample size | sales | total assets | debt/ total assets |
|---|---|---|---|---|---|
| Metal Products Industry | 1985 | 62 | 5345 | 3667 | .685 |
| Printing Industry | 1984 | 96 | 2075 | 1602 | .729 |
| Retail trade in leather goods | 1984 | 43 | 452 | 348 | .519 |
| Clothes shops | 1983 | 137 | 806 | 472 | .417 |
| Domestic utensils shops | 1983 | 112 | 537 | 504 | .438 |
| Retail trade in paint | 1982 | 64 | 497 | 228 | .578 |
| Record shops | 1982 | 67 | 698 | 306 | .699 |
| Shoemakers | 1982 | 68 | 169 | 105 | .461 |
| Domestic decoration shops | 1981 | 155 | 1200 | 869 | .534 |
| Structural Engineering | 1981 | 112 | 2422 | 1307 | .596 |
| Stationer's shops | 1980 | 116 | 1139 | 607 | .480 |
| Electro-technical retailers | 1980 | 81 | 1160 | 482 | .619 |
| Grocery stores | 1979 | 207 | 2163 | 373 | .743 |
| Clothes shops | 1979 | 100 | 1035 | 458 | .472 |

## 7.3 Hypotheses and variables

Assets

In the literature, comparatively little attention is paid to the factors that determine the firm's asset requirements. In the theory of finance, the firm's productive assets are usually assumed to be already in place, in order not to confuse financing decisions with investment decisions. In the practical literature attention is likewise scant. However, the asset categories on the firm's balance sheet represent the value of factors of production, so a number of expected determinants of the size and structure of the assets can be adopted from production theory. In neoclassical production theory, the relation between outputs and various inputs (such as labour, capital, energy, etc.) is represented in a production function. The mathematical properties of a production function can be chosen to reflect the characteristics of a production process as regards the existence of scale effects, the degree of substitutability of the factors of production, etc. Usually, a certain degree of substitutability is assumed, which means that a particular level of production can be achieved with different combinations of factors of production. Under a set of "neoclassical" assumptions, the optimal quantity of a factor of production can be determined, depending on the scale of the firm's operations, the price of the product, the prices of all factors of production and the used quantities of other factors of production. On the basis of such relations, two effects can be studied and empirically quantified, viz. scale effects, i.e. the presence of constant, decreasing or increasing returns to scale, and substitution effects, i.e. the effects of changes in the relative prices of factors of production on the optimal quantities of these factors.

Of course, the derivation of relations for asset categories from a fully specified theoretical model would be beyond the scope of this study. Instead, a more pragmatic approach is followed, in which elements from the neoclassical production theory are used to specify relations for asset categories in the data at hand. In these relations, asset categories are related to a set of explanatory variables, which are (partly) adopted from neoclassical production theory. The asset categories represent the "quantities" of factors of production we seek to explain and the explanatory variables can be associated with the two effects mentioned above, viz. a scale effect and a substitution effect, as well as heterogeneity effects. The latter refer to differences in products or in factors of production, which are assumed to be homogeneous in the theory.

It should also be noted, however, that the use of the neoclassical approach to the analysis of production in retailing, to which most of our samples refer, has been severely criticized. For instance, Nooteboom (1980) argues that, among other things, substituting capital for labour changes the nature of shops, so that they have to be classified in a different shoptype, e.g. a grocery shop becomes a supermarket. This effect will occur more frequently as shoptypes are defined more narrowly and it would make substitution effects more relevant for analyses of shoptypes than for analyses of individual firms within a shoptype. However, in this study we use a more general concept of capital (i.e. assets) than Nooteboom, where shop surface is used. Moreover, in the pragmatic approach followed here we do not want to exclude a priori the possibility of substitution effects, so elements from neoclassical production theory are adopted in the analyses of assets.

Scale, or the size of the firm's operations, is measured as sales size. In the analysis of assets, sales is the most obvious size variable because it is an output measure, comparable across firms and shoptypes. In the analysis of production, other size variables also occur, e.g. value added or gross profit, which may have a more direct relation with production than sales. However, in the analysis of financial structure sales is a preferable size variable, because sales are directly related to asset categories as accounts receivables, which are usually ignored in production analysis. Sales have to be financed, so to speak. The amount of assets will, of course, increase with sales size, but the increase is likely to be less than proportional, i.e. negative scale effects in the ratio assets-to-sales are expected.

The substitution effect between labour and capital is expressed in the relations for assets by including the share of labour costs in total costs as an explanatory variable. For a given size of operations, a high share of labour costs leads, according to production theory, to a lower amount of assets. The reverse is also true. A real substitution effect cannot be specified for the data at hand, but the specification used here allows characteristics of the production process to be incorporated in the analysis. The shares of the various cost categories can be regarded as resultants of decisions, that are made in the past on the basis of the prices at that time. So the specification used here facilitates the de facto choice of production process to be taken into account in the relations for asset categories. Although "production process effect" might be a better name for the effect specified here, the more common term substitution effect is used.

Heterogeneity effects occur if different firms within an industry are engaged in different activities (supply different products) or use different factors of production. For both elements, three variables are included in the analyses.

- The first variable to express differences in activities is the <u>assortment composition.</u> For each sample, two or three different categories of products or activities are distinguished. The choice of categories is motivated by expected differences in asset requirements but, in view of the wide variety of different products, no hypotheses as to the exact nature of these differences are formulated. Table 7.2 gives an overview of the categories used. In most samples, assortment composition is measured as the share of the product categories in total sales (i.e. total sales equals the sum of the categories). In the samples provided by Rabobank Nederland, assortment composition is measured by a dummy variable for industrial classification.

Table 7.2 Assortment composition of the samples

| Sample | Assortment group 1 | Assortment group 2 | Assortment group 3 |
|---|---|---|---|
| Metal Products Industry* | doors, windows greenhouses | other metal products | – |
| Printing Industry* | offset print shops | general print shops | remaining print shops |
| Retail trade in leather goods | ladies'&men's bags | remaining sales | – |
| Clothes shops | clothes | fashion–and other articles | – |
| Domestic utensils shops | glass, china, etc. | domestic & garden utensils | – |
| Retail trade in paint | paint, wall–paper, etc. | tools, ironware, etc. | – |
| Record shops | records, tapes, & cassettes | equipment & accessories | – |
| Shoemakers | repair | retail sales | – |
| Domestic decoration shops | furniture | floorcoverings & curtains | handicraft |
| Structural Engineering | light con–structions | heavy constructions | – |
| Stationer's shops | stationery | office equipment | remaining sales |
| Electro–technical retailers | electrical appliances | remaining sales | – |
| Grocery stores | foods | non–foods & beverages | – |
| Clothes shops | clothes | fashion– and other articles | – |

*Note table 7.2:* * *means dummy variables are used to indicate assortment.*

- The second indicator for differences in activities is the gross margin, which is included to reflect the service level. The general service level of shops has many dimensions: depth and width of the assortment, spaciousness of the shop, assistance and advice offered by the shop–assistants, etc. The gross margin is used to approximate this complex concept, because all elements of service have to be compensated for in the retail price. More service is, of course, expected to be associated with the use of more resources including assets.
- The third variable for differences in activities is a dummy variable reflecting the firm's assortment policy (dummy deep assortment). This variable takes the value one for firms with a stated policy to carry a particularly wide or deep assortment as an element of their marketing mix. For the other firms its value is zero.
- The first variable reflecting heterogeneity in factors of production is the price level of housing costs (i.e. the costs per square metre). This variable is likely to have several effects. The price level of housing costs determines to a considerable extend the costs of keeping goods in stock, so a high price level will lead to small warehouses, low inventories and, hence, low current assets. Similarly, a high price level of housing costs will negatively influence the volume component of buildings and fixtures, but the effect on the value per unit is of course positive, so that the combined effect on fixed assets is hard to predict. In most samples, an imputed rent is used for firms that do not pay current market–price rents, either because they own the buildings or because they rent it from themselves or relatives. This procedure makes the variable a good indicator of the commercial quality of the point of location, (which influences the relation between fixed assets and sales because on good locations high sales can be obtained), but a bad indicator of the costs of keeping goods is in stock. So this variable's influence is limited to fixed assets and, since the commercial quality is primarily relevant for retailing, to the retailing samples, while for current assets the following variable is used.
- The second variable reflecting heterogeneity in factors of production is the interest rate, calculated as the amount of interest paid divided by total debt. Firms facing high interest rates, e.g. because they cannot offer collateral or because they have exhausted their normal debt capacity, are likely to curtail investments. So current assets, which are readily adjustable, will be kept low. The interest rate is frequently not available because in most samples only the balance of interest paid and interest

received is included.

- The last heterogeneity refers to the ownership of buildings which, in the samples at hand, is measured by a dummy variable (dummy owning buildings). The importance of this variable is obvious, since buildings only appear on the balance sheet if they are firm property.

Debt

As regards the right hand side of the balance sheet, hypotheses can be formulated on the basis of the determinants of financial structure that are described in the previous chapters and that are summarized in table 6.1. This table also contains a list of empirical proxy variables and expected effects. The hypotheses in table 6.1 and the literature on which they are based are extensively discussed in the previous chapters, so that here a short description will suffice.

Since all expected effects in table 6.1 are formulated in terms of more or less debt, the obvious choice of dependent variable is debt, leaving equity to be implicitly estimated as the balancing item of assets and debt. Separate relations for long and short term debt are used to allow for the different effects on both categories expressed in some hypotheses. The hypotheses are grouped into the categories size effects, tax effects, collateral/agency effects and miscellaneous effects.

In the analysis of debt categories, total assets is used as size variable in a specification that on the one hand resembles the commonly used debt ratios (long and short term debt–to–total assets), and that on the other hand is flexible enough to allow for scale effects in these ratios. Positive scale effects are expected, i.e. the ratio of debt–to–total assets is expected to increase with total assets. For lack of more precise measures, the tax advantage of debt approximated by the percentage depreciation charges in total costs.

Collateral and agency effects are usually associated with the same firm characteristics, as a high collateral value implies less agency costs. To approximate these combined effects the following five variables are used. First of all, the structure of the assets is used in the analysis, expressed as the percentage fixed assets in total assets. This readily available variable is used rather than an arbitrary split up of fixed assets in collateralizable and non–collateralizable assets. In the literature, asset structure is frequently used as an indication of collateral value. Secondly, inventory turnover is included in the analysis. This variable is not frequently encountered in the literature but, in retailing, inventories are an important element of assets. In chapter 5 it was mentioned that the nature of the inventories determines whether or not they are accepted by banks as collateral for loans. Inventory turnover is likely to reflect the liquidity of and, thus, the security offered by the inventories. It should be noted, however, that the inventory turnover reflects the characteristics of the principal items in the assortment. Within most industries (shoptypes) these characteristics will vary considerably less than between shoptypes, so that the influence of inventory turnover may be limited. The third variable for collateral and agency effects is a dummy variable (dummy mortgages) which reflects whether or not long term loans are collateralized by mortgages. The fourth variable is also a dummy (dummy other collaterals) which reflects whether or not other securities are given to collateralize loans. These securities include fiduciary transfers of property of the inventories, cession of claims and private collaterals. The expected, obviously positive, influence of these dummies on debt will need no elucidation. The fifth and last variable referring to collateral and agency effects is also a dummy (dummy limited liability), which summarizes the firm's legal form. This variable takes the value one for firms with limited liability[4] and is zero otherwise. Since limited liability reduces the possibilities to collect claims or necessitates additional costs to bypass these limitations in the debt contract, a negative influence on debt is expected.

The category miscellaneous effects comprises seven variables which are discussed below, together with there expected influences.

- the dummy family loans, which measures the incidence of loans from friends and relatives, is expected to have a positive effect on debt since such loans signal information to potential debtholders about the quality of the firm's investments (based on "inside information" from the entrepreneur).
- Recent (substantial) investments are expected to be associated with (at least temporarily) high debt ratios because small firms may not be able to generate enough funds internally to finance investments with the current proportions of debt and equity. In most samples, investments are measured in money amounts, but in three cases (domestic utensils shops, domestic decoration shops and grocery stores) only a dummy variable is available.
- Investment plans, on the other hand, are hypothesized to be associated with low debt ratios, because earnings are retained in advance in order to accumulate reserve borrowing power. Investment plans are measured with a dummy variable.

------

[4]*In Dutch: N.V.'s en B.V.'s.*

- Growth is measured as the average annual sales growth over the last 2, 3 or 4 years, depending on the availability of data. Myers' growth effect specifies a positive influence on short term and negative influence on long term debt, while the signal effect of growth on debt is positive.
- Co-ownership implies agency problems associated with equity and can cause various conflicts of interest between the partners regarding e.g. paying dividends or retaining earnings. Hence, co-owned firms are expected to rely more heavily on debt. Co-ownership is measured by the number of entrepreneurs in the firm.
- The age of the entrepreneur is included in the analyses, although in the practical literature the influence of this variable is motivated by simple correlations and the theoretical motivation of this variable, if any, is established in a rather roundabout way However, the analyses provide an opportunity to see to what extent older entrepreneurs do indeed use less debt.
- The age of the firm is included to the reflect the life cycle effects that are frequently mentioned in the literature. Since no precise definitions of life cycles in terms of years of existence are provided, age is included as such. In the samples at hand, the very small, starting firms are excluded, so a high age can be expected to be associated with high debt ratios.
- Finally, Myers' pecking order effect is represented by the composition of equity, i.e. the share of retained earnings in total equity which reflects the cumulative profitability over an extended period. According to the pecking order effect, observed debt ratios reflect the cumulative requirement for external financing over an extended period which will be negatively related to profitability.

## 7.4 Specification and estimation results

The specification, which is used to test the hypotheses formulated in the previous sub-section, is chosen to conform to the following two requirements. First, the specification should adequately reflect the theoretical arguments. This means that e.g. arbitrary groupings to express a continuous influence should be avoided as far as possible. Second, the specification should establish a link with the practice of financial analyses. To accomplish this, a specification is used that can be considered a more general equivalent of ratios that are commonly used in financial statement analysis. The generality should allow the various hypothesized effects to be adequately expressed, while the equivalence to ratios should facilitate an easy application of the results.

In the previous sub-section, 15 different hypotheses are formulated which, combination with the 14 samples, would lead to 210 different tests. However, the actual number of tests is lower, because none of the samples contains all variables necessary to test the 15 hypotheses. Also, in the large amount of data at hand it is almost inevitable that some pairs of variables correlate so strongly that only one of them can be included in the analysis. Further, some variables appear to have too little empirical variation to be included in any analysis. For instance, in the sample of shoemakers none of the firms is co-owned, while only one of them has limited liability. Nevertheless, a large number of hypotheses remains and they are tested with the following specification :

$$(7.1)\quad Y_{ik} = \beta_1 + \sum_{h \in R_k} \beta_{2h} S_{hik} \left[ \prod_{j \in V_k} X_{ij}^{\beta_{3j}} \right] exp \sum_{l \in W_k} \beta_{4l} Z_{il} + u_i$$

in which :

| | |
|---|---|
| $Y$ | =the variable to be explained (fixed and current assets and short long term debt respectively), |
| $S$ | =size variable (described below), |
| $X,Z$ | =explanatory variables described in the text, |
| $R_k$ | =the set of indices for size variables $S_h$ used in the explanation of variable $Y_k$ (sales in 2 or 3 groups for current assets, sales and sales x dummy own buildings for fixed assets and total assets for debt categories), |
| $V_k$ | =the set of indices for the explanatory variables $X_j$ used in the explanation of variable $Y_k$, |
| $W_k$ | =the set of indices for the explanatory variables $Z_l$ used in the explanation of variable $Y_k$, |
| $i$ | =index for firm |
| $k$ | =index for the variable to be explained ($k = 1,2,3,4$) |
| $\beta$ | =coefficients to be estimated, |
| $u$ | =disturbance term. |

The explanatory variables which are strictly positive are specified as $X$; these variables have been transformed, i.e. they are divided by their sample average value. The explanatory variables that can be negative or zero are specified in an exponent of the constant $e$. The disturbance term $u$ is assumed to be an independently distributed stochastic variable with zero expectation and constant variance[5]. A multiplicative specification is chosen for $X$ and $Z$ because such a specification allows for interaction between variables. Scale effects are accounted for by the intercept ($\beta_1$). When this coefficient appears to be different from zero, the ratio $Y_k/S_k$ is not a constant but scale–dependent, i.e. dependent on size variable $S$. This scale effect becomes clear after rewriting $Y_k = \beta_1 + \beta_2 S_k$ as $Y_k/S_k = \beta_1/S_k + \beta_2$. The ratio $Y_k/S_k$ consists of two parts, the scale–dependent term $\beta_1/S_k$ and the scale–independent term $\beta_2$. This specification facilitates an easy interpretation and application of the results. For instance, if $Y_k$ is short term debt, then $\beta_2$ is equivalent to the ratio of short term debt to total assets if all other coefficients in the equation do not differ significantly from zero. If scale effects appear ($\beta_1$ significantly different from zero), $\beta_2$ can be interpreted as the scale–adjusted ratio of short term debt to total assets[6]. For the full equation, $\beta_2$ represents the scale–adjusted ratio for shoptypes with average values of $X_i$ and zero values for $Z_j$.

Equation (7.1) is estimated for the four variables to be explained (current and fixed assets and short and long term debt) in separate analyses, using the data of the fourteen samples. Equation (7.1) is not linear and the estimation results are produced by a non linear least squares fit using Marquardt's (1963) algorithm. No analytical expression is available for the standard errors of the estimated coefficients, they are computed numerically within the algorithm. It is assumed that the estimated coefficients are asymptotically normally distributed. The results are presented in table 7.3 through 7.6. These tables also shows the precise specification of (7.1) in terms of variable composition and whether variables are specified as X or Z[7]. A summarizing and, occasionally, somewhat telegraphic discussion of these voluminous estimation results is presented below.

-------

[5] *Inspection of scatter plots of the variables to be explained and explanatory variables revealed some heteroscedasticity which is, however, ignored here.*

[6] *An alternative specification of scale effects is* $Y = \beta_1 S^{\beta_2}....$, *which can be easily estimated after logarithmic transformation. In fact, this specification could be more suited for the analysis of debt categories, since it implies zero debt at zero total assets. On the other hand, scale effects in assets are likely to take the form of a threshold (cf. Nooteboom, 1980), which can be adequately expressed in the intercept of specification (7.1). Because scale effects are expected to be more important for assets than for debt, and in order to maintain a uniform specification throughout the analyses, specification (7.1) is used.*

[7] *The note accompanying these tables is included at the end of this chapter.*

Table 7.3 Estimation results current assets

| | | Metal products industry | Printing industry | Retail leather goods | Clothes shops 1983 | Domestic utensils shops |
|---|---|---|---|---|---|---|
| Intercept | $\hat{\beta}_1$ | -73.494 | 86.739 | 51.427 | 63.672 | 155.860 |
| | | (79.597)* | (26.899) | (17.513) | (16.872) | (25.736) |
| Sales group 1 | $\hat{\beta}_{21}$ | .659 | .628 | .338 | .247 | .148 |
| | | (.055) | (.070) | (.106) | (.025) | (.062) |
| Sales group 2 | $\hat{\beta}_{22}$ | -.149 | -.311 | .360 | .380 | .261 |
| | | (.032) | (.067) | (.115) | (.107) | (.066) |
| Sales group 3 | $\hat{\beta}_{23}$ | – | -.182 | – | – | – |
| | | | (.063) | | | |
| Gross margin | $\hat{\beta}_{31}$ | .260 | -.444 | 1.146 | -.249 | .674 |
| | | (.076) | (.140) | (.792)* | (.320)* | (.702) |
| % labour costs | $\hat{\beta}_{32}$ | -.050 | .570 | .146 | -.003 | .931 |
| | | (.115)* | (.187) | (.112)* | (.046)* | (.338) |
| Interest rate | $\hat{\beta}_{41}$ | -8.775 | -9.624 | -3.516 | -4.105 | .497 |
| | | (1.884) | (1.237) | (1.761) | (1.314) | (1.112)* |
| Dummy deep assortment | $\hat{\beta}_{42}$ | – | – | – | – | .357 |
| | | | | | | (.155) |
| Degree of fit | $\rho^2$ | .985 | .952 | .763 | .588 | .582 |

Table 7.4 Estimation results fixed assets

| | | Metal products industry | Printing industry | Retail leather goods | Clothes shops 1983 | Domestic utensils shops |
|---|---|---|---|---|---|---|
| Intercept | $\hat{\beta}_1$ | 36.197 | 94.945 | 55.964 | 71.080 | -4.063 |
| | | (153.030)* | (77.593)* | (31.253)* | (22.345) | (22.473)* |
| Sales | $\hat{\beta}_{21}$ | .250 | .636 | -.011 | .013 | .099 |
| | | (.030) | (.042) | (.028)* | (.024)* | (.047) |
| Sales x dummy own buildings | $\hat{\beta}_{22}$ | .128 | .360 | .672 | .474 | .461 |
| | | (.027) | (.049) | (.092) | (.023) | (.051) |
| Gross margin | $\hat{\beta}_{31}$ | .357 | 1.407 | -6.965 | 1.677 | 1.310 |
| | | (.153) | (.256) | (1.791) | (.329) | (.510) |
| % labour costs | $\hat{\beta}_{32}$ | -.592 | -1.280 | -.005 | -.074 | -.555 |
| | | (.117) | (.187) | (.105)* | (.052)* | (.094) |
| Housing costs | $\hat{\beta}_{33}$ | – | – | .453 | .251 | – |
| | | | | (.234) | (.111) | |
| Degree of fit | $\rho^2$ | .909 | .773 | .656 | .789 | .624 |

*Note tables 7.3 and 7.4 : see end of this chapter*

Table 7.3 (continued) Estimation results current assets

| | | Retail trade in paint | Record shops | Shoe-makers | Domestic decoration shops | Structural enginee-ring |
|---|---|---|---|---|---|---|
| Intercept | $\hat{\beta}_1$ | 74.241 | 46.001 | 22.995 | -.641 | 35.268 |
| | | (11.718) | (24.090) | (4.870) | (31.074)* | (42.137)* |
| Sales group 1 | $\hat{\beta}_{21}$ | .075 | .174 | .032 | .326 | .345 |
| | | (.032) | (.050) | (.078)* | (.023) | (.025) |
| Sales group 2 | $\hat{\beta}_{22}$ | .217 | .275 | .700 | .354 | .273 |
| | | (.062) | (.052) | (.327) | (.053) | (.016) |
| Sales group 3 | $\hat{\beta}_{23}$ | - | - | - | 1.255 | - |
| | | | | | (.413) | |
| Gross margin | $\hat{\beta}_{31}$ | .047 | .707 | - | -.222 | -.215 |
| | | (.719)* | (.294) | | (.258)* | (.108) |
| % labour costs | $\hat{\beta}_{32}$ | .399 | -.006 | -.081 | -.080 | .344 |
| | | (.232) | (.110)* | (.147)* | (.098)* | (.241)* |
| Interest rate | $\hat{\beta}_{41}$ | 1.382 | .404 | -11.020 | - | - |
| | | (1.495)* | (1.725)* | (5.294) | | |
| Dummy deep assortment | $\hat{\beta}_{42}$ | - | .085 | - | -.039 | - |
| | | | (.117)* | | (.051)* | |
| Degree of fit | $\rho^2$ | .628 | .746 | .259 | .771 | .893 |

Table 7.4 (continued) Estimation results fixed assets

| | | Retail trade in paint | Record shops | Shoe-makers | Domestic decoration shops | Structural enginee-ring |
|---|---|---|---|---|---|---|
| Intercept | $\hat{\beta}_1$ | 51.666 | 28.529 | 47.572 | 50.681 | 237.093 |
| | | (14.657) | (12.795) | (9.015) | (57.913)* | (62.141) |
| Sales | $\hat{\beta}_{21}$ | -.003 | .019 | -.008 | .020 | -.008 |
| | | (.017)* | (.013)* | (.042)* | (.066)* | (.040)* |
| Sales x dummy own buildings | $\hat{\beta}_{22}$ | .394 | .336 | .384 | .426 | .176 |
| | | (.074) | (.030) | (.083) | (.055) | (.039) |
| Gross margin | $\hat{\beta}_{31}$ | 3.859 | -1.067 | 6.530 | .343 | 1.256 |
| | | (1.559) | (.415) | (4.761)* | (.563)* | (.417) |
| % labour costs | $\hat{\beta}_{32}$ | -.181 | .084 | -.087 | -.347 | -.519 |
| | | (.088) | (.124)* | (.104)* | (.098) | (.559)* |
| Housing costs | $\hat{\beta}_{33}$ | .967 | 1.477 | .533 | .003 | - |
| | | (.685)* | (.230) | (.252) | (.186)* | |
| Degree of fit | $\rho^2$ | .611 | .801 | .473 | .591 | .561 |

*Note tables 7.3 and 7.4 : see end of this chapter.*

Table 7.3 (continued) Estimation results current assets

| | | Stationer's shops | Electrotechnical retailers | Grocery stores | Clothes shops 1979 |
|---|---|---|---|---|---|
| Intercept | $\hat{\beta}_1$ | 33.194 | 59.675 | 41.189 | 60.770 |
| | | (23.835)* | (18.022) | ( 7.745) | (21.165) |
| Sales group 1 | $\hat{\beta}_{21}$ | .217 | .218 | .038 | .160 |
| | | (.031) | (.029) | (.007) | (.021) |
| Sales group 2 | $\hat{\beta}_{22}$ | .347 | .233 | .150 | .345 |
| | | (.021) | (.025) | (.032) | (.050) |
| Sales group 3 | $\hat{\beta}_{23}$ | .210 | – | – | – |
| | | (.035) | | | |
| Gross margin | $\hat{\beta}_{31}$ | .413 | .510 | .037 | .488 |
| | | (.205) | (.174) | (.289)* | (.333)* |
| % labour costs | $\hat{\beta}_{32}$ | -.127 | -.087 | .165 | -.101 |
| | | (.090)* | (.042) | (.198)* | (.085)* |
| Interest rate | $\hat{\beta}_{41}$ | – | – | – | – |
| Dummy deep assortment | $\hat{\beta}_{42}$ | .151 | -.123 | – | – |
| | | (.058) | (.090)* | | |
| Degree of fit | $\rho^2$ | .889 | .823 | .731 | .652 |

Table 7.4 (continued) Estimation results fixed assets

| | | Stationer's shops | Electrotechnical retailers | Grocery stores | Clothes shops 1979 |
|---|---|---|---|---|---|
| Intercept | $\hat{\beta}_1$ | 42.576 | 14.374 | 18.619 | 37.130 |
| | | (25.309) | (25.562)* | (20.840)* | (29.608)* |
| Sales | $\hat{\beta}_{21}$ | .014 | .039 | .047 | .056 |
| | | (.023)* | (.019) | (.010) | (.030) |
| Sales x dummy own buildings | $\hat{\beta}_{22}$ | .303 | .227 | .087 | .201 |
| | | (.021) | (.021) | (.008) | (.020) |
| Gross margin | $\hat{\beta}_{31}$ | 1.085 | -.134 | -.855 | -.756 |
| | | (.311) | (.372)* | (.397) | (.462)* |
| % labour costs | $\hat{\beta}_{32}$ | -.285 | .117 | -.325 | -.155 |
| | | (.090) | (.104)* | (.127) | (.086) |
| Housing costs | $\hat{\beta}_{33}$ | -.118 | -.133 | -.347 | – |
| | | (.107)* | (.146)* | (.157) | |
| Degree of fit | $\rho^2$ | .783 | .680 | .573 | .643 |

*Note tables* 7.3 *and* 7.4 : *see end of this chapter.*

Assets

Regarding the relation between the asset categories and the size variable used in analyses of assets, sales, the following remarks can be made:

- Scale effects are more important for current than fixed assets. The expected, significantly positive intercepts are found in 10 of the 14 analyses of current assets, and in 6 of the 14 analyses of fixed asset. Significantly negative intercepts are not encountered.
- Although no formal test is applied, differences in assortment composition appear to be associated with different current asset requirements in most samples. Note that the negative coefficients of the sales in group 2 and 3 in the metal products and printing industries are caused by the fact that, in these samples, assortment composition is indicated by a dummy variable, so that these results do not imply negative asset requirements.
- Not surprisingly, owning buildings is associated with significantly higher fixed assets requirement than renting buildings is in all samples. In eight samples, the coefficient of sales is insignificant, probably because buildings constitute the major part of fixed assets and because in some samples the majority of the firms own their buildings, so that the remaining fixed assets can be very small.

The gross margin, reflecting the general service level, appears to have ambiguous effects: the hypothesis that more service is associated with the use of more assets is accepted in 12 of the 28 cases (5 in current assets, 7 in fixed assets); it is rejected in 5 cases (2 in current assets, 3 in fixed assets), while the remaining 11 coefficients are insignificant.

Substitution effects, reflected by the share of labour costs in total costs, are primarily relevant for fixed assets as, with hindsight, could have been expected. In 8 of the 14 samples, a high share of labour is associated with low investments in fixed assets, while in the remaining 6 samples insignificant effects are found (4 of which are negative). In the analyses of current assets, the effect of the share of labour costs is predominantly insignificant (10 samples), although it is significantly positive in 3 and negative in 1 sample(s). Surprisingly, no substitution effects are found in the structural engineering industry, although non-retailing industries are expected to have more opportunities to substitute capital for labour than retailing industries.

The interest rate, indicating the costs of keeping goods in stock, is indeed associated with low current assets in 5 of the 8 samples in which this variable is available; in the remaining samples an insignificant effect is found. On the other hand, the dummy deep assortment, which approximates assortment policy, has a significantly positive effect on current assets in only 2 of the 5 samples in which its influence is tested. Perhaps positive effects on inventories, to which this variable refers, are cancelled out by negative effects on remaining current assets. Finally, housing costs per square meter have a significantly positive influence on fixed assets in 4 out of 9 samples, a significantly negative influence in 1 and an insignificant influence in the remaining 4 samples. Since the effect of this variable is the balance of a negative effect on the "volume" component of fixed assets and a positive effect on the "price per unit" component, the mixed results obtained with this variable are not very surprising.

Summarizing, important determinants of current assets appear to be size (scale effects), assortment composition and interest rate, while service (gross margin), substitution effects and assortment policy are less important. Fixed assets primarily depend on the ownership of buildings, the substitution of capital for labour and service (gross margin), while scale-effects and housing costs per square metre have a weaker influence.

Debt

In the samples at hand, scale effects appear to play a minor role in the determination of the debt ratio since insignificant intercepts are obtained in 24 of the 28 analyses. Surprisingly, the 4 significant intercepts are all positive, while negative ones are expected. So, if there are scale effects, the (short term) debt ratio decreases with size. The coefficients of the size variable (total assets) are significantly positive in all but one case (the shoemaker's short term debt).

The tax advantage of debt is approximated by the percentage depreciation charges in the total costs, which is hypothesized to be accompanied by the use of little debt. This hypothesis is rejected more often than it is accepted (in 6 versus 5 cases), but in the large majority of cases (17) the effect of this variable is insignificant. So, little support is found for the tax shelter hypothesis.

The percentage fixed assets in total assets is one of the proxy variables for the collateral value of assets, so that a large fixed asset share is expected to be associated with comparatively high debt ratios. However, assets structure appears to influence the maturity structure of debt rather than the overall debt ratio. In the analyses of short term debt 9 significantly negative coefficients are found and in the analyses of long term debt 10 significantly positive coefficients are obtained; the remaining

Table 7.5 Estimation results short term debt

| | | Metal products industry | Printing industry | Retail leather goods | Clothes shops 1983 | Domestic utensils shops |
|---|---|---|---|---|---|---|
| Intercept | $\hat{\beta}_1$ | −35.471 | −34.018 | 22.648 | 44.338 | 39.785 |
| | | (89.230)* | (37.612)* | (17.411)* | (16.353) | (14.041) |
| Total assets | $\hat{\beta}_2$ | .487 | .376 | .194 | .065 | .089 |
| | | (.147) | (.078) | (.055) | (.032) | (.030) |
| % depreciation costs | $\hat{\beta}_{31}$ | .149 | −.091 | .279 | −.120 | .337 |
| | | (.055) | (.063)* | (.181)* | (.130)* | (.208)* |
| % fixed assets | $\hat{\beta}_{32}$ | −.437 | −.381 | −.342 | .105 | −.121 |
| | | (.079) | (.068) | (.166) | (.166)* | (.120)* |
| Inventory turnover | $\hat{\beta}_{33}$ | .073 | −.078 | – | −.916 | −.256 |
| | | (.022) | (.032) | | (.327) | (.298)* |
| Growth | $\hat{\beta}_{34}$ | – | – | −3.859 | −4.796 | −5.374 |
| | | | | (3.169)* | (1.979) | (3.570)* |
| Co-ownership | $\hat{\beta}_{35}$ | .137 | .055 | – | .455 | −1.213 |
| | | (.086)* | (.076)* | | (.325)* | (.682) |
| Age entrepreneur | $\hat{\beta}_{36}$ | .200 | .272 | .172 | −.987 | −1.290 |
| | | (.149)* | (.080) | (.877)* | (.898)* | (.724) |
| Age business | $\hat{\beta}_{37}$ | – | – | −.026 | .151 | .275 |
| | | | | (.210)* | (.255)* | (.191)* |
| Dummy other collaterals | $\hat{\beta}_{41}$ | – | – | – | – | – |
| Dummy limited liability | $\hat{\beta}_{42}$ | −.076 | .371 | .307 | .323 | .702 |
| | | (.284)* | (.194) | (.363)* | (.229)* | (.290) |
| Dummy family loans | $\hat{\beta}_{43}$ | −.246 | −2.279 | – | – | – |
| | | (.568)* | (.775) | | | |
| Recent investments (dummy) | $\hat{\beta}_{44}$ | – | – | −.258 | .066 | – |
| | | | | (.193)* | (.045)* | |
| Dummy investment plans | $\hat{\beta}_{45}$ | – | – | −.113 | .312 | – |
| | | | | (.304)* | (.208)* | |
| % retained earnings | $\hat{\beta}_{46}$ | .354 | −.762 | – | – | – |
| | | (.175) | (.106) | | | |
| Degree of fit | $\rho^2$ | .988 | .948 | .405 | .466 | .483 |

*Note table 7.5 : see end of this chapter.*

Table 7.5 (continued) Estimation results short term debt

| | | Retail trade in paint | Record shops | Shoe-makers | Domestic decoration shops | Structural enginee-ring |
|---|---|---|---|---|---|---|
| Intercept | $\hat{\beta}_1$ | 44.931 | 25.167 | 12.826 | 14.342 | 35.546 |
| | | (14.401) | (18.723)* | ( 2.530) | (21.119)* | (32.824)* |
| Total assets | $\hat{\beta}_2$ | .055 | .275 | .010 | .091 | .646 |
| | | (.052) | (.082) | (.015)* | (.026) | (.083) |
| % depreciation costs | $\hat{\beta}_{31}$ | -.553 | .007 | -2.602 | .359 | -.026 |
| | | (.376)* | (.117)* | (1.184) | (.092) | (.054)* |
| % fixed assets | $\hat{\beta}_{32}$ | -.453 | -.371 | -.772 | -.282 | -.209 |
| | | (.299)* | (.095) | (1.662)* | (.052) | (.048) |
| Inventory turnover | $\hat{\beta}_{33}$ | -1.004 | .076 | - | .127 | -.210 |
| | | (.857)* | (.200)* | | (.094)* | (.053) |
| Growth | $\hat{\beta}_{34}$ | 2.880 | -6.856 | -44.746 | -2.144 | -1.483 |
| | | (3.407)* | (2.206) | (27.183) | ( .544) | ( .415) |
| Co-ownership | $\hat{\beta}_{35}$ | - | -.023 | - | -.063 | -.264 |
| | | | (.379)* | | (.115)* | (.085) |
| Age entrepreneur | $\hat{\beta}_{36}$ | 1.777 | .212 | -10.244 | .271 | - |
| | | (1.943)* | (.558)* | (6.415)* | ( .226)* | |
| Age business | $\hat{\beta}_{37}$ | -.486 | -.172 | 3.615 | - | - |
| | | (.512)* | (.126)* | (2.079) | | |
| Dummy other collaterals | $\hat{\beta}_{41}$ | - | - | - | 1.091 | - |
| | | | | | (.127) | |
| Dummy limited liability | $\hat{\beta}_{42}$ | - | .084 | - | .106 | -.537 |
| | | | (.185)* | | (.093)* | (.089) |
| Dummy family loans | $\hat{\beta}_{43}$ | - | - | - | - | - |
| Recent invest-ments (dummy) | $\hat{\beta}_{44}$ | - | .078 | 1.306 | .606 | - |
| | | | (.067)* | (.752) | (.165) | |
| Dummy invest-ment plans | $\hat{\beta}_{45}$ | - | .309 | - | -.232 | - |
| | | | (.207)* | | (.120) | |
| % retained earnings | $\hat{\beta}_{46}$ | - | - | - | - | - |
| Degree of fit | $\rho^2$ | .187 | .677 | .175 | .832 | .879 |

*Note table 7.5 : see end of this chapter.*

Table 7.5 (continued) Estimation results short term debt

| | | Stationer's shops | Electrotechnical retailers | Grocery stores | Clothes shops 1979 |
|---|---|---|---|---|---|
| Intercept | $\hat{\beta}_1$ | 7.303 | -7.346 | 10.056 | 35.069 |
| | | (16.819)* | (27.056)* | (13.178)* | (22.140)* |
| Total assets | $\hat{\beta}_2$ | .245 | .472 | .375 | .140 |
| | | (.038) | (.057) | (.045) | (.042) |
| % depreciation costs | $\hat{\beta}_{31}$ | .132 | .264 | .127 | -.309 |
| | | (.094)* | (.129) | (.097)* | (.171) |
| % fixed assets | $\hat{\beta}_{32}$ | -.185 | -.313 | -.630 | -.145 |
| | | (.064) | (.065) | (.104) | (.113)* |
| Inventory turnover | $\hat{\beta}_{33}$ | -.002 | -.128 | -.064 | -.802 |
| | | (.129)* | (.174)* | (.110)* | (.253) |
| Growth | $\hat{\beta}_{34}$ | 1.548 | -3.360 | -1.411 | 4.682 |
| | | (.464) | (1.109) | (.758) | (1.751) |
| Co-ownership | $\hat{\beta}_{35}$ | - | - | .090 | - |
| | | | | (.161)* | |
| Age entrepreneur | $\hat{\beta}_{36}$ | - | - | .012 | .230 |
| | | | | (.214)* | (.553)* |
| Age business | $\hat{\beta}_{37}$ | .105 | -.156 | -.144 | -.114 |
| | | (.066)* | (.065) | (.046) | (.139)* |
| Dummy other collaterals | $\hat{\beta}_{41}$ | .343 | - | - | - |
| | | (.110) | | | |
| Dummy limited liability | $\hat{\beta}_{42}$ | .153 | - | .043 | - |
| | | (.117)* | | (.092)* | |
| Dummy family loans | $\hat{\beta}_{43}$ | .112 | - | - | - |
| | | (.095)* | | | |
| Recent investments (dummy) | $\hat{\beta}_{44}$ | -.060 | - | .171 | - |
| | | (.025) | | (.079) | |
| Dummy investment plans | $\hat{\beta}_{45}$ | -.363 | - | -.055 | |
| | | (.152) | | (.085)* | - |
| % retained earnings | | - | - | - | - |
| Degree of fit | $\rho^2$ | .841 | .597 | .599 | .359 |

*Note table 7.5 : see end of this chapter.*

Table 7.6 Estimation results long term debt

| | | Metal products industry | Printing industry | Retail leather goods | Clothes shops 1983 | Domestic utensils shops |
|---|---|---|---|---|---|---|
| Intercept | $\hat{\beta}_1$ | 65.062 | 41.678 | -25.942 | 24.553 | -6.881 |
| | | (54.365)* | (29.779)* | (26.094)* | ( 8.680)* | (14.042)* |
| Total assets | $\hat{\beta}_2$ | .141 | .237 | .440 | .043 | .212 |
| | | (.048) | (.047) | (.106) | (.021) | (.031) |
| % depreciation costs | $\hat{\beta}_{31}$ | -.246 | -.552 | -.255 | .016 | -.162 |
| | | (.071) | (.069) | (.124) | (.111)* | (.172)* |
| % fixed assets | $\hat{\beta}_{32}$ | 1.069 | 1.336 | -.086 | 1.650 | 1.058 |
| | | (.227) | (.215) | (.156)* | (.483) | (.178) |
| Inventory turnover | $\hat{\beta}_{33}$ | -.104 | .027 | – | -1.392 | -.458 |
| | | (.053) | (.039)* | | (.320) | (.135) |
| Growth | $\hat{\beta}_{34}$ | – | – | -.997 | -1.636 | 3.708 |
| | | | | (3.693)* | (1.193)* | (1.595) |
| Co-ownership | $\hat{\beta}_{35}$ | .058 | -.427 | – | -1.676 | -.382 |
| | | (.160)* | (.119) | | (.760) | (.173) |
| Age entrepreneur | $\hat{\beta}_{36}$ | .160 | -.402 | -1.370 | -2.718 | -.668 |
| | | (.317)* | (.119) | (.772) | (.833) | (.266) |
| Age business | $\hat{\beta}_{36}$ | – | – | .417 | .766 | .008 |
| | | | | (.303)* | (.257) | (.064)* |
| Dummy mortgages | $\hat{\beta}_{41}$ | – | – | – | – | – |
| Dummy other collaterals | $\hat{\beta}_{42}$ | – | – | – | – | – |
| Dummy limited liability | $\hat{\beta}_{43}$ | -.079 | .051 | -.558 | -.102 | -.175 |
| | | (.276)* | (.168)* | (.515)* | (.126)* | (.125)* |
| Dummy family loans | $\hat{\beta}_{44}$ | .541 | 2.542 | – | – | – |
| | | (.817)* | (.588) | | | |
| Recent investments (dummy) | $\hat{\beta}_{45}$ | – | – | -.293 | .162 | – |
| | | | | (.181)* | (.036) | |
| Dummy investment plans | $\hat{\beta}_{46}$ | – | – | -.508 | .079 | – |
| | | | | (.228) | (.174)* | |
| % retained earnings | $\hat{\beta}_{47}$ | .100 | -.425 | – | – | – |
| | | (.270)* | (.110) | | | |
| Degree of fit | $\rho^2$ | .958 | .887 | .785 | .709 | .821 |

*Note table 7.6 : see end of this chapter.*

Table 7.6 (continued) Estimation results long term debt

| | | Retail trade in paint | Record shops | Shoe-makers | Domestic decoration shops | Structural enginee-ring |
|---|---|---|---|---|---|---|
| Intercept | $\hat{\beta}_1$ | 2.792 | -17.536 | -6.364 | 2.102 | 22.931 |
| | | (10.669)* | (11.495)* | ( 5.497)* | (29.262)* | (39.962)* |
| Total assets | $\hat{\beta}_2$ | .136 | .246 | .282 | .169 | .114 |
| | | (.061) | (.055) | (.066) | (.052) | (.058) |
| % depreciation costs | $\hat{\beta}_{31}$ | -.266 | .501 | -.160 | -.067 | -.305 |
| | | (.216)* | (.111) | (.163)* | (.164)* | (.216)* |
| % fixed assets | $\hat{\beta}_{32}$ | 1.881 | .442 | 1.860 | .238 | .919 |
| | | (.636) | (.120) | (.536) | (.214)* | (.300) |
| Inventory turnover | $\hat{\beta}_{33}$ | .283 | -.316 | .034 | -.211 | .330 |
| | | (.220)* | (.124) | (.122)* | (.139)* | (.166) |
| Growth | $\hat{\beta}_{34}$ | 1.484 | 2.085 | -.566 | 2.179 | 2.613 |
| | | (3.495)* | (1.638)* | (1.054) | ( .771) | ( .778) |
| Co-ownership | $\hat{\beta}_{35}$ | - | -.958 | - | -.404 | .714 |
| | | | (.482) | | (.160) | (.314) |
| Age entrepreneur | $\hat{\beta}_{36}$ | -.722 | -.498 | -.904 | -1.093 | - |
| | | (.423) | (.392)* | (.280) | (.352) | |
| Age business | $\hat{\beta}_{37}$ | .238 | -.277 | -.120 | - | - |
| | | (.175)* | (.101) | (.057) | | |
| Dummy mortgages | $\hat{\beta}_{41}$ | - | 1.062 | - | .266 | - |
| | | | (.210) | | (.133) | |
| Dummy other collaterals | $\hat{\beta}_{42}$ | - | - | - | .483 | - |
| | | | | | (.130) | |
| Dummy limited liability | $\hat{\beta}_{43}$ | - | -.072 | - | -.348 | -.308 |
| | | | (.111)* | | (.101) | (.378)* |
| Dummy family loans | $\hat{\beta}_{44}$ | - | - | - | - | - |
| Recent investments (dummy) | $\hat{\beta}_{45}$ | - | -.153 | -.062 | .295 | - |
| | | | (.057) | (.041)* | (.192)* | |
| Dummy investments plans | $\hat{\beta}_{46}$ | - | -.241 | - | -.235 | - |
| | | | (.131) | | (.147)* | |
| % retained earnings | $\hat{\beta}_{47}$ | - | - | - | - | - |
| Degree of fit | $\rho^2$ | .787 | .914 | .830 | .719 | .361 |

*Note table 7.6 : see end of this chapter.*

Table 7.6 (continued) Estimation results long term debt

| | | Sta- tioner's shops | Electro- technical retailers | Grocery stores | Clothes shops 1979 |
|---|---|---|---|---|---|
| Intercept | $\hat{\beta}_1$ | -6.642 | 7.126 | 1.013 | -15.046 |
| | | (11.995)* | (13.604)* | (12.736)* | (17.140)* |
| Total assets | $\hat{\beta}_2$ | .085 | .092 | .173 | .135 |
| | | (.023) | (.029) | (.031) | (.037) |
| % depreciation costs | $\hat{\beta}_{31}$ | .476 | .095 | -.230 | .206 |
| | | (.126) | (.186)* | (.108) | (.137)* |
| % fixed assets | $\hat{\beta}_{32}$ | .215 | 1.183 | .277 | 1.449 |
| | | (.171)* | (.323) | (.266)* | (.437) |
| Inventory turnover | $\hat{\beta}_{33}$ | -.549 | .131 | -.393 | -.741 |
| | | (.158) | (.262)* | (.119) | (.201) |
| Growth | $\hat{\beta}_{34}$ | .522 | -.408 | 2.721 | 3.865 |
| | | ( .580)* | (1.040)* | ( .699) | (1.497) |
| Co-ownership | $\hat{\beta}_{35}$ | - | - | -2.067 | - |
| | | | | (1.286)* | |
| Age entrepreneur | $\hat{\beta}_{36}$ | - | - | -.580 | -.087 |
| | | | | (.254) | (.491)* |
| Age business | $\hat{\beta}_{37}$ | .003 | -.655 | -.354 | -.108 |
| | | (.073)* | (.111) | (.057) | (.117)* |
| Dummy mortgages | $\hat{\beta}_{41}$ | .550 | - | .401 | - |
| | | (.124) | | (.127) | |
| Dummy other collaterals | $\hat{\beta}_{42}$ | .187 | - | - | - |
| | | (.113) | | | |
| Dummy limited liability | $\hat{\beta}_{43}$ | -.094 | - | -.201 | - |
| | | (.108)* | | (.087) | |
| Dummy family loans | $\hat{\beta}_{44}$ | .532 | - | - | - |
| | | (.115) | | | |
| Recent investments (dummy) | $\hat{\beta}_{45}$ | .036 | - | .145 | - |
| | | (.028)* | | (.089)* | |
| Dummy investment plans | $\hat{\beta}_{46}$ | .244 | - | .047 | - |
| | | (.114) | | (.084)* | |
| % retained earnings | $\hat{\beta}_{47}$ | - | - | - | - |
| Degree of fit | $\rho^2$ | .838 | .767 | .732 | .560 |

*Note table* 7.6 : *see end of this chapter.*

5 respectively 4 coefficients are insignificant. These results provide only limited support for the collateral value hypothesis in the modern theory of finance, they rather reflect the traditional rule of finance proclaiming the synchronization of the terms of investing and financing (the "golden balance sheet rule" as it is called in The Netherlands). Although this traditional rule of finance has been severely criticized, the results obtained here conform the research by Donckels et al. (1987), which indicates that synchronizing the terms of investing and financing still plays an important role in the financial policy of SME's. Comparable results, although far less pronounced, are obtained with inventory turnover, which has a negative influence on the long term debt ratio in 7 samples, while the expected positive influence is found in only one sample (the remaining 5 samples show insignificant coefficients). This variable's influence on short term debt is predominantly insignificant (7 insignificant, 1 significant positive and 4 significant negative coefficients are obtained). The possibility of weaker results with this variable in the analysis of samples of individual firms was anticipated in the previous sub-section.

Two hypotheses regarding the effect of growth on financial structure are formulated in the literature, viz. Myers' growth effect, which specifies a positive influence on short term debt and a negative influence on long term debt, and the signal effect of growth which is unambiguously positive. For Myers' growth effect, no support is found at all, as a positive effect on short term debt and a negative effect on long term debt are found in none of the samples. The signal effect of growth is only supported in the clothes shops 1979. Instead, growth appears to have a negative influence an short term debt in 7 of the 12 samples in which growth is available, and a positive influence on long term debt in 5 samples.

Co-owned firms are hypothesized to rely more heavily on debt because of the agency problems of equity associated with co-ownership and because conflicts of interest can be avoided by using more debt (rather than using more equity). This effect is not found in any of the samples. Only one significantly positive coefficient is encountered (long term debt, structural engineering), but this sample shows a significantly negative coefficient in the analysis of short term debt. Contrary to the hypothesis, the influence of co-ownership appears to be negative, particularly on long term debt.

Older entrepreneurs are reported to use less debt and, somewhat surprisingly, this effect is indeed found, at least as far as long term debt is concerned: 8 out of 11 coefficients are significantly negative. For short term debt the age of the entrepreneur is largely irrelevant.

Life cycle effects, specifying that as firms grow older they use more debt, are hardly found in the data. In the analyses of both long and short term debt only 1 out of 10 coefficients is significantly positive. Negative coefficients are encountered more frequently (2 respectively 4 times) but the majority of coefficients is insignificant (12 out of 20).

The dummy mortgages, which is only included in the analyses of long term debt, has the anticipated significantly positive effect in all 4 samples in which its influence is tested. The dummy other collaterals, which is included in the analyses of both debt categories, also has the expected positive effect in the two samples in which it is available. The effects of these two variables appear to be quite strong and it is a pity that they are only available in a minority of the samples.

The effect of limited liability appears to be limited, for 15 of the 20 coefficients are insignificant. The expected negative effects on debt are only found in 3 cases (2 long term, 1 short term debt), while two significantly positive effects, rejecting the hypothesis, are also found (both short term debt). The limited importance of legal form probably reflects the general practice of banks to ask security for loans from either business or private sources, regardless the legal form (cf. the discussion in section 2.4.4).

Family loans are expected to increase debt capacity by signalling positive information about the firms investment projects to the capital markets. As regards long term debt, this effect is found in 2 of the 3 samples in which family loans are recorded; for short term debt the hypothesis is rejected in one sample (and accepted in none).

Recent investments and conversely, investment plans can influence financial structure if firms cannot generate enough equity to finance investments with the current proportions of debt and equity. Hence, investments lead to (temporarily) high debt ratios and investment plans to low debt ratios. This effect of investments is found in 4 out of 14 cases; three of these 4 refer to short term debt. The hypothesis is rejected in 2 cases and the remaining 8 cases show insignificant results. The hypothesis regarding investment plans is rejected in 1 case and accepted in 4 while the remaining 7 cases are insignificant.

Finally, the hypothesis that debt ratios are negatively related to cumulative profitability, which is approximated by the % retained earnings in total equity, is tested in two samples. In the printing industry, the hypothesis is accepted; in the metal products industry, it is rejected for short term debt and for long term debt the result is insignificant.

The multitude of estimation results presented above may have a slightly dazing effect and in order to provide some overview, a summary of the estimation results is presented in table 7.7. From this table, important determinants of long and short term debt (besides, of course, size) appear to be asset structure and inventory turnover (although these variables reflect traditional financing rules rather than theoretical considerations) and the dummies reflecting collateral securities (mortgages and other collaterals). To a lesser degree, the entrepreneur's age, recent investments, investment plans and family loans are also relevant. Scale effects, tax shelter effects, life cycle effects and the effect of limited liability appear to be of little importance, while the hypotheses regarding growth and co-ownership are mainly rejected. The (scant) empirical evidence on Myers' pecking order effect is mixed[8].

Reviewing the results in terms of industries it can be observed that the two industries which have the lowest asset requirements (retail trade in paint and shoemakers) also show the lowest degree of fit in the analyses of short term debt. In the sample of shoemakers, even the size variable failed to produce a significant coefficient. Apparently, the very small amounts of short term debt in these samples are hardly subject to the primarily economic considerations tested here. The degree of fit of the analyses of long term debt in these samples is high and a similar remark applies to the present analyses in general.

-------

8 *Notes tables* 7.3 *through* 7.6 :

- *estimated standard errors are printed beneath the coefficients*;
- *an asterisk * is printed next to the standard error if for the coefficient* $\hat{\beta}$, $|\hat{\beta}| < 1.64\ \hat{\sigma}(\hat{\beta})$ *obtains, i.e. if the coefficient is not significantly different from zero at a* 10% *level of significance.*
- *the square of the correlation coefficient between the variable to be explained and its estimation* ($\rho^2$) *is used to measure the degree of fit.*

Table 7.7 Summary of estimation results debt

| | | Hypothesis | Number of tests | Accepted | Rejected | Inconclusive |
|---|---|---|---|---|---|---|
| Intercept | $\hat{\beta}_1$ | – | 28 | 0 | 4 | 24 |
| Total assets | $\hat{\beta}_2$ | + | 28 | 27 | 0 | 1 |
| % depreciation costs | $\hat{\beta}_{31}$ | – | 28 | 5 | 6 | 17 |
| % fixed assets | $\hat{\beta}_{32}$ | + | 28 | 10 | 9 | 9 |
| Inventory turnover | $\hat{\beta}_{33}$ | + | 25 | 2 | 11 | 12 |
| Growth (Signal) | $\hat{\beta}_{34}$ | + | 24 | 7 | 8 | 9 |
| Co-ownership | $\hat{\beta}_{35}$ | + | 16 | 1 | 7 | 8 |
| Age entrepreneur | $\hat{\beta}_{36}$ | – | 22 | 9 | 1 | 12 |
| Age business | $\hat{\beta}_{37}$ | + | 20 | 2 | 6 | 12 |
| Dummy mortgages | $\hat{\beta}_{41}$ | + | 4 | 4 | 0 | 0 |
| Dummy other collaterals | $\hat{\beta}_{42}$ | + | 4 | 4 | 0 | 0 |
| Dummy limited liability | $\hat{\beta}_{43}$ | – | 20 | 3 | 2 | 15 |
| Dummy family loans | $\hat{\beta}_{44}$ | + | 6 | 2 | 1 | 3 |
| Recent investments (dummy) | $\hat{\beta}_{45}$ | + | 14 | 4 | 2 | 8 |
| Dummy investment plans | $\hat{\beta}_{46}$ | – | 12 | 4 | 1 | 7 |
| % retained earnings | $\hat{\beta}_{47}$ | – | 4 | 2 | 1 | 1 |

*Note table* 7.7: *a test is classified as "accepted" if the corresponding coefficient is significantly different from zero in the direction specified by the hypothesis; it is classified as "rejected" if the corresponding coefficient is significantly different from zero in the opposite direction as the one specified by the hypothesis and "inclusive" if the coefficient is insignificant.*

## 8.1 Introduction

In this chapter, two databases of industry averages in the retail trade are analysed. The information in these databases consists of averaged financial statements for narrowly defined shoptypes and is publicly available: the information has been published by the collectors for purposes such as interfirm comparisons[1]. Analysing data on this level of aggregation offers several advantages over the obvious alternative of analysing data of individual firms within an industry. First of all, analysing industry averages allows the inclusion of variables that differ strongly between, but hardly within shoptypes, such as the characteristics of the principal items in the assortment (e.g. Titman's durable effect). Conversely, using aggregated data also allows abstraction from variables that vary widely within, but probably much less between shoptypes, such as differences in valuation caused by differences in recency of investments (i.e. the effects of cumulative depreciation). Another example is the influence of the personal wealth of the entrepreneur on the financial behaviour of the firm. This type of variables typically cannot be modelled for lack of available data. Finally, industry averages are published more often than information of individual firms, so using the former increases the number of available data sources.

The data have a panel character: information on shoptypes is available for a number of years[2]. Data of this type can be analysed in several ways. For instance, the information available in each year can be treated as a separate cross section sample. Alternatively, the data of each shoptype can be used as a separate time series, although in the data at hand the number of years is somewhat limited for that. Additionally, the data can be pooled an analysed in several ways. Pooling offers the opportunity to use the panel character of the data and techniques to analyse panel data are more and more reported on (e.g. Hsiao, 1986), reflecting the increasing availability of panel data.

In this study, the data are pooled but a full exploitation of the

-------

[1] *The data sources are described in the next subsection.*

[2] *The databases do not have a complete panel character: for some shoptypes information is available for less years than it is for others.*

panel character of the data is relegated to subsequent study: the empirical analyses are confined to an exploratory investigation of the databases. In an earlier paper (Spronk and Van der Wijst, 1988), these data are analysed in a series of separate cross–section studies for each year available in the data. This is a methodologically sound first approach to analysing these data, but an important drawback of this research design is that, in combination with the data, it produces a large number of outcomes (in the 1988 paper, 96 separate regression analyses are reported on). This multitude of results may serve to reveal, with minimal restrictions on the outcomes, development patterns over time. However, if no clear patterns emerge, which seems to be the case, the sheer bulk of the outcomes thwarts their interpretation and evaluation. Therefore, in this study the information per database of the different years is pooled and the thus combined time series and cross section data are analysed in two ways. First, in section 8.3 financial structure is analysed in terms of money amounts (representing the value of asset and debt categories), as is usually the case in financial statement analysis. Second, in section 8.4 the data are also analysed from a somewhat different perspective. Analysing financial structure in terms of money values (i.e. the amounts that appear in the firm's financial statement) may, to a certain extent, obscure the behaviour which induces the year–to–year changes, because the observed values are the resultants of behaviour over a number of years. This is particularly true if the yearly changes are small compared with the observed amounts, as is the case with the average data at hand. In order to expose this behaviour more sharply, this study also comprises an analysis of changes, i.e. first differences between the observed amounts. The analysis of first differences is confined to debt categories.

## 8.2 Data

The data used in this chapter refer to the retail trade in the Netherlands and in Western Germany and were primarily collected to facilitate interfirm comparisons by entrepreneurs and small business consultants. The data are based on information of individual firms, but the unit of observation used here is the "industry average", i.e. the averaged data for narrowly defined shoptypes such as supermarkets and shoe shops. The Dutch data were collected by the Research Institute for Small and Medium–Sized Business in the Netherlands. The data cover a ten year period (1976 through 1985) for 16 shoptypes and a six year period (1980 through 1985) for an additional 23 shoptypes. In this study, only the former 16 shoptypes are used. The average number of firms per shoptype ranges between 50 and 80. The data comprise fairly detailed financial statements and additional information on labour volume, legal form, shop surface and number of branches. For this study, the industry (i.e. shoptype) averages were calculated from the individual data, but similar information on a subset of the variables is published regularly[3].

The German data were collected by the "Institut für Handelsforschung an der Universität zu Köln" in two separate series of surveys. The first series consists of regular inquiries into the financial situation of the retail trade, made every alternate year over the period 1955 through 1969 and the years 1973 and 1977. Because beginning and ending balance sheets were collected, an uninterrupted row of ending balance sheets over the period 1954 through 1969 can be constructed, as well as ending balance sheets for the years 1972, 1973, 1976 and 1977. The corresponding sales sizes are only available every other year. The research outcomes, i.e. average balance sheets for some thirty shoptypes, have been published by the Cologne research institute. The balance sheet data for this study have been taken from two ten year summaries and a periodical, both brought out by the research institute mentioned above[4].

The second series of surveys consists of yearly inquiries into the turnover, costs, margins and profits of the retail trade over the period 1949 through 1985. In these surveys, information on income statements is collected and published for largely the same shoptypes as in the balance sheet survey. The income statement data for this study have been taken from three ten year summaries brought out by the Cologne institute mentioned earlier[5]. Both series of surveys are claimed to be representative of the German retail trade

-------

[3] *See e.g. "Bedrijfssignalering: bedrijfsresultaten* 1984" *and "Bedrijfs-signalering: balanspositie* 1984", *published by "Economisch Instituut voor het Midden- en Kleinbedrijf", Zoetermeer (Holland),* 1985.

[4] *"Die Vermögens-und Kapitalsituation des Einzelhandels in dem Jahrzehnt* 1954 *bis* 1963" *respectively* "1964 *bis* 1973" *published in* 1966 *respectively* 1975 *in the series "Sonderhefte der Mitteilungen des Instituts für Handelsforschung an der Universität zu Köln" as Sonderheft* 17 *respectively* 25 *by Westdeutscher Verlag, Köln respectively Verlag Otto Schwartz, Göttingen and: "Die Vermögens-und Kapitalsituation des Einzelhandels in den Jahren* 1976 *und* 1977" *in: "Mitteilungen des Instituts für Handelsforschung an der Universität zu Köln", Jahrg.* 31, *no.* 4, *April* 1979.

[5] *"Umsatz, Kosten, Spannen und Gewinn des Einzelhandels in der Bundesrepublik Deutschland in dem Jahrzehnt* 1949 *bis* 1958" *respectively* "1959 *bis* 1968" *respectively* "1969 *bis* 1978" *published in* 1962 *respectively* 1970 *respectively* 1980 *in the series "Schriften zur Handelsforschung" as Schrift no.*22 *respectively* 44 *respectively* 63 *by the Westdeutscher Verlag, Köln,* (*n.* 22 *and* 44) *and Verlag Otto Schwartz, Göttingen* (*no.* 63).

for all variables except size variables (sales size, total assets). From the publications, sales size (and sample size) can be seen to differ between both surveys, but the only ratio they have in common, the inventory turnover, correlates almost perfectly between the surveys (simple correlation coefficient always >.94, usually >.97). Hence, it seems safe to use ratios from the income statement survey in the analysis of balance sheet data. The combined surveys contain concise balance sheets and sales sizes (which can be transformed into German marks), margins and costs as percentages of sales (which cannot be transformed into German marks because they are unweighted averages), and information on labour volume, shop surface, sales growth, sales on credit and paid or imputed rent per square metre shop surface. This information is available for 28 shoptypes over the entire period and for 14 other shoptypes over a part of the period. Only the former 28 shoptypes are used in this study. A more detailed description of the data can be found in the publications referred to earlier.

## 8.3 Analysis of financial structure

### 8.3.1 *Hypotheses and variables*

Assets

The analyses of assets in this chapter are based on the same theoretical considerations as the previous chapter's analyses. These considerations refer to some broad categories of determinants of the size and structure of the assets, which are largely adopted from neoclassical production theory. Recall that these categories are:

- scale effects, which refer to the way assets increase with size, i.e. proportionally or more or less than proportionally;
- substitution effects, which refer to the production process, i.e. the used combination of labour and capital (assets), and
- heterogeneity effects, which refer to differences in the products sold or means of production used.

Most of the variables, which are used to measure or approximate these effects, are similar to the ones described in the previous chapter, so that they need little further discussion. The following variables are used:

Scale, or the size of the firm's operations,is measured as sales size. Again, negative scale effects in the ratio of assets-to-sales are anticipated i.e. economies of scales in asset requirements are expected.

The substitution effect between labour and capital is expressed in the relations for assets by including the share of labour costs in total costs as an explanatory variable. According to Nooteboom's argument, this variable may have a stronger effect in the present analysis of shoptype averages than in had in the previous chapter. Of course, negative effects on assets are expected.

Heterogeneity effects may occur because different firms or shoptypes may be engaged in different activities, i.e. supply a different product (or "bundle of services"), or use different factors of production. For both elements, two variables are included in the relations for assets:

- the first indicator of the activities is a dummy variable for shoptypes selling mainly foods and perishable goods (dummy foods and perishables); this variable has the value one for shoptypes as supermarkets, greengrocers, butchers and florists and is zero for the other shoptypes. Because of their high speed of turnover, shoptypes selling foods and perishables are likely to have comparatively low asset requirements;
- the second indicator of the activities is the gross margin, which is included to reflect the service level. As is already argued in the previous chapter, more service is, of course, expected to be associated with the use of more resources including assets;
- the first variable reflecting heterogeneity in factors of production is the price level of housing costs (i.e. the costs per square meter). A high price level of housing costs is expected to be associated with low current assets, while the effect on fixed assets is hard to predict;
- the last heterogeneity refers to the ownership of buildings, as opposed to renting buildings, which is clearly of importance since buildings only appear on the balance sheet if they are firm property. The ownership relations in a shoptype are approximated by the depreciation charges on buildings as a fraction of the total housing costs (i.e. depreciations, rent and remaining housing costs). In the German data, an imputed rent is used for firms owning their buildings, so the variable for ownership of buildings is only available for the Dutch data.

In addition to the three effects discussed above, two variables are included in the specification to express the time dimension present in the data. The first variable is the average annual growth rate in size[6]. This variable is included to reflect the continuous process of scaling up, which has been a dominant characteristic of the retail trade over the entire post-war period. This process, however, is not likely to have affected all shoptypes in the same degree. For instance, today's supermarkets have evolved from the small grocery shops that were common a few decades ago, but clothes shops have changed much less. To capture the long term nature of this process, the average growth rate is used rather than the year-to-year changes. Scaling up is likely to affect many aspects of the way shops are operated. Although the effects on current and fixed assets may be different, the overall effect on assets is expected to be positive because scaling up is generally associated with the introduction of more capital intensive selling techniques, such as self-service. Hence, scaling up and the substitution effect discussed earlier can overlap, but since scaling up represents a wider concept, both variables may be assumed to reflect different aspects of the labour-capital combination. Finally, a time trend is included to reflect the effects of variables that influence the amounts of assets over time, but that are not included in the analysis.

Debt

As regards the right hand side of the balance sheet, hypotheses can again be formulated on the basis of the determinants of financial structure that are summarized in table 6.1. The hypotheses, which are tested on the basis of the pooled data, can be grouped in four categories, viz. size effects, collateral effects, tax effects and miscellaneous effects.

In the analysis of debt categories, total assets is again used as size variable; positive scale-effects are expected, i.e. the ratio of debt-to-total assets is expected to increase with total assets.

The collateral value of assets is approximated by three different variables. The first two of these, the structure of the assets, expressed as the percentage fixed assets in total assets, and inventory turnover are already introduced in the previous chapter. The third and last variable referring to collateral security is a dummy variable for shoptypes selling primarily goods which are subject to fashion, such as clothes, soft

-------

[6] *The growth rate for shoptype* $i = \sqrt[T]{size_{t=T}/size_{t=1}}$ , $t = 1,2,..T$. *Note that technically this growth rate implies that information of the last year is used in the analysis of the first year. However, this variable is used to approximate a phenomenon that has been present in the shoptypes over the entire period.*

furnishings, sportswear, etc. (dummy fashion). These shoptypes are generally considered to be more risky, because fashion is hard to predict and the lines of supply are long[7], while the inventories and the shop-equipment may bring in only a fraction of their purchasing price in a forced sale.

The tax advantage of debt is approximated by the percentage depreciation charges in total costs. This variable is chosen because it is commonly used in the literature and no other information on non-debt tax shields is available.

Myers' growth effect and the signal effect of growth are tested with the average annual growth rate in size, which is calculated in the same manner as the growth variable in the analysis of assets, but with total assets as size variable. Titman's durable effect is captured with a dummy variable for industry classification (dummy durables) which takes the value one for shoptypes selling primarily durables and the value zero otherwise. Finally, a time trend is included in the relations for debt categories in order to reflect the influence of not included variables.

### 8.3.2 *Specification and estimation results*

The specification, which is used to test the hypotheses formulated in the previous sub-section, is similar to the one used in the previous chapter. In addition to the general requirements formulated in chapter 7, the specification should also take into account that the data at hand have a panel character. This means that variables can be either shoptype-varying time-invariant variables, or time-varying shoptype-invariant or shoptype-time varying variables (see e.g. Hsiao, 1985). This distinction is of importance since most variables tend to change over time, so that the "time series effects" may cancel out or even reverse the cross section effects and vice versa. Hence, it is necessary to specify exactly what type of effect is intended. In the empirical analyses in this subsection, cross section effects are emphasized; time series effects are analysed in section 8.4.

Panel data offer the possibility to control for effects of shoptype-specific and/or time-specific omitted variables, by e.g. including dummy variables for each shoptype and/or period (see Hsiao, 1986). This procedure is not followed here. In the earlier analysis of the data (Spronk and Van der Wijst, 1988), time effects were found to be of minor importance so that, in the present analysis, they are assumed to be accounted for by the time trend.

-------

[7] *For instance, the summer collection of clothes shops is usually purchased at the beginning of winter.*

Instead of a separate dummy for each shoptype, this analysis includes a smaller number of dummies (the dummies foods & perishables, fashion and durables). These dummies are not likely to eliminate completely the effects of shoptype-specific omitted variables. However, this procedure offers the advantages of a simpler specification, interpretable dummy variables and a greater applicability. The following specification is used:

$$(8.1)\quad Y_{ikt} = \beta_1 + \beta_2 S_{ikt} \left[ \prod_{j \in V_k} X_{ijt}^{\beta_{3j}} \right] exp \sum_{l \in W_k} \beta_{4l} Z_{ilt} + u_{it}$$

in which :

$Y$ =the variable to be explained (fixed and current assets and short long term debt respectively),

$S$ =size variable (sales for assets, total assets for debt),

$X,Z$ =explanatory variables described in the text,

$V_k$ =the set of indices for the explanatory variables $X_j$ used in the explanation of variable $Y_k$,

$W_k$ =the set of indices for the explanatory variables $Z_l$ used in the explanation of variable $Y_k$,

$i$ =index for shoptype

$k$ =index for the variable to be explained ($k = 1,2,3,4$)

$t$ =index for time

$\beta$ =coefficients to be estimated,

$u$ =disturbance term.

The variables to be explained and the size variables are shoptype–time varying variables. With the exception of size growth, the same applies to the explanatory variables which are strictly positive and, hence, are specified as $X$; these variables have been transformed, i.e. they are divided by their average value in each year. This procedure preserves the cross sectional differences, but removes the changes in the averages values of these variables over time. Because size growth is measured as an average annual growth rate over the entire period, it is a shoptype varying time invariant variable. Its average value is close to 1 (1.03). All dummy variables are, of course, also time invariant, while the time trend is a time varying shoptype invariant variable. The explanatory variables that can be negative or zero are specified in an exponent of the constant $e$. The time trend is likewise specified to reflect an autonomous growth rate. The dummy variables are specified in the same manner because they can be zero. The disturbance term $u$ is assumed to be a stochastic variable with an intertemporally independent distribution, zero mean and constant variance. However, since the effects of shoptype–specific omitted variables is only partially accounted for by dummy variables, this assumption may not be met[8]. A multiplicative specification is again chosen for $X$ and $Z$ because such a specification allows for interaction between variables.

Equation (8.1) is estimated for the four variables to be explained in separate analyses, using the data referring to the two countries. Equation (8.1) is not linear and the estimation results are produced by the non linear least squares fit using Marquardt's (1963) algorithm. The estimation results are presented in the tables 8.1 and 8.2, which also show the precise specification of (8.1) in terms of variable composition. These tables are briefly discussed below.

The results regarding assets (table 8.1) show, that the amounts of assets are, of course, positively related to sales, the size variable in this analysis. In both countries, the coefficient for fixed assets is considerably lower than the one for current assets. The expected scale effects are only found for fixed assets in the Dutch data and for current assets in the German data.

The share of labour costs in total costs intends to capture the effects of substitution processes, so a negative coefficient is expected. This effect is indeed found in three of the four analyses; only for fixed assets in the Dutch data an insignificant (positive) coefficient is obtained. The substitution effect appears to be particularly strong in the analysis of fixed assets, German data.

The four heterogeneities included in the analysis of assets largely appear to have the hypothesized effects. The dummy variable for shoptypes selling primarily foods and perishable products (dummy foods and perishables) has a significantly negative effect in all cases, as expected. The gross margin has a significantly positive influence on assets, conformably to the hypothesis, in three of the four cases and an insignificantly negative influence in one case. The price level of housing costs is expected to have a negative effect on current assets, while no hypothesis is formulated as regards fixed assets. The effects of this variable appear to be negative in all cases and significantly so in all but one, viz. current assets, Dutch data. The variable reflecting the ownership of buildings is associated with a significantly higher amount of fixed assets, in accordance with the

-------

[8] *Inspection of the residuals of* (8.1), *printed in an I times T matrix, indeed revealed some correlation per shoptype over time, but this is ignored here.*

hypothesis, for the Dutch data. This variable is not available in the German data.

The two variables which are included to express the time dimension in the data, appear to have very different effects. The <u>average annual growth rate</u> in size is included to reflect the process of scaling up, so a positive effect on assets is expected. However, for the Dutch data a positive effect

Table 8.1 Estimation results current and fixed assets

| | | Dutch data | | German data | |
|---|---|---|---|---|---|
| | | fixed assets | current assets | fixed assets | current assets |
| Intercept | $\hat{\beta}_1$ | 25.3373 | -15.7513 | -.1638 | 22.9398 |
| | | (6.9040) | (11.8032)* | (3.1115)* | (6.0163) |
| Sales | $\hat{\beta}_2$ | .1411 | .3750 | .1190 | .2403 |
| | | (.0115) | (.0271) | (.0056) | (.0089) |
| Housing costs per square m. | $\hat{\beta}_{31}$ | -.2077 | -.0034 | -.6176 | -.1772 |
| | | (.0260) | (.0263)* | (.0349) | (.0222) |
| Gross margin | $\hat{\beta}_{32}$ | .6523 | 1.0545 | .8946 | -.1444 |
| | | (.1021) | (.1325) | (.1039) | (.0766)* |
| % labour costs | $\hat{\beta}_{33}$ | .0723 | -.9952 | -1.2608 | -.2762 |
| | | (.1419)* | (.1048) | (.0839) | (.0636) |
| Size growth | $\hat{\beta}_{34}$ | 2.0587 | -2.8072 | -5.6707 | .6232 |
| | | (.8166) | (.8360) | (.6315) | (.4412)* |
| Renting /owning buildings | $\hat{\beta}_{35}$ | .3190 | - | - | - |
| | | (.0377) | | | |
| Dummy foods & perishables | $\hat{\beta}_{41}$ | -.4847 | -.9719 | -.6211 | -1.3683 |
| | | (.0370) | (.0576) | (.1292) | (.1707) |
| Time | $\hat{\beta}_{42}$ | .0477 | .0116 | -.0030 | -.0036 |
| | | (.0045) | (.0052) | (.0016)* | (.0012) |
| n. of observations | | 160 | 160 | 560 | 560 |
| Degree of fit | $\rho^2$ | .932 | .854 | .923 | .951 |

*Notes table 8.1:*
- *estimated standard errors are printed beneath the coefficients;*
- *an asterisk * is printed next to the standard error if for the coefficient $\hat{\beta}$, $|\hat{\beta}| < 1.96\ \hat{\sigma}(\hat{\beta})$ obtains, i.e. if the coefficient is not significantly different from zero at a 5% level of significance.*
- *the square of the correlation coefficient between the variable to be explained and its estimation ($\rho^2$) is used to measure the degree of fit.*

on fixed assets and a negative effect on current assets are found, while for the German data opposite results are obtained. The effects appear to be quite strong, except for current assets, German data, which show an insignificant coefficient. The time trend, for which no hypothesis is formulated, is positive in the Dutch data and negative in the German data and significant in three of the four cases. In view of the fact that there are hardly any indications that the retail trades in Germany and in The Netherlands have developed along different lines, there is no ready explanation for these opposite results.

The results regarding debt (table 8.2) show, that the coefficients of the size variable, total assets, are significantly positive, as could be expected. In both countries, the coefficient for short term debt is somewhat larger than the one for long term debt. Both coefficients in the Dutch data are considerably larger than their counterparts in the German data. Positive scale effects are expected, i.e. the ratio of debt to total assets is expected to increase with total assets. Indeed, all intercepts appear to be negative, which causes an increase in the ratio as size increases. However, only two of the four intercepts are significantly negative, evenly distributed over short and long term debt and both countries.

Asset structure, expressed as the percentage fixed assets in total assets, is one of the three proxy variables for the collateral value of assets, so that a large fixed assets share is expected to be associated with comparatively high debt ratios. Significantly positive coefficients for this variable are indeed found in the analyses of long term debt. For short term debt, on the other hand, significantly negative coefficients are found. Asset structure thus appears to influence the maturity structure of debt more than the overall debt ratio. Again, these results provide only limited support for the collateral value hypothesis in the theory of finance, they rather reflect the traditional rule of finance proclaiming the synchronization of the terms of investing and financing (the "golden balance sheet rule" as it is called in The Netherlands). Much the same results, although of opposite sign, are obtained with the inventory turnover, the second proxy variable for the collateral value of assets. This variable produces significantly positive coefficients in the analyses of short term debt and significantly negative ones in the case of long term debt. As was the case in the previous chapter, the results obtained here conform the research by Donckels et al. (1987), which indicates that synchronizing the terms of investing and financing still plays an important role in the financial policy of SME's.

The third and last variable referring to collateral security is the dummy variable for fashion. Because fashion is associated with riskiness, a negative influence on debt is expected. All coefficients of this variable are indeed negative, but they are only significant in the Dutch data.

The tax advantage of debt is approximated by the percentage depreciation charges in total costs. Since depreciation charges may have a limiting effect on the tax deductibility of interest charges, large non-debt

Table 8.2 Estimation results short and long term debt

| | | Dutch data | | German data | |
|---|---|---|---|---|---|
| | | short term debt | long term debt | short term debt | long term debt |
| Intercept | $\hat{\beta}_1$ | -9.8803 | -6.5976 | -2.9441 | -4.2698 |
| | | (4.6540) | (4.5800)* | (2.0940)* | (1.0919) |
| Total assets | $\hat{\beta}_2$ | .4824 | .4303 | .2515 | .1894 |
| | | (.0275) | (.0284) | (.0109) | (.0064) |
| Asset structure | $\hat{\beta}_{31}$ | -.8100 | .8255 | -.4412 | .7058 |
| (fixed/tot.ass.) | | (.1062) | (.1186) | (.0435) | (.0368) |
| Inventory | $\hat{\beta}_{32}$ | .2508 | -.1543 | .3030 | -.2439 |
| turnover | | (.0327) | (.0335) | (.0224) | (.0221) |
| Tax shields | $\hat{\beta}_{33}$ | -.5348 | -.2031 | .3736 | -.1203 |
| (depr.charges) | | (.1040) | (.1034) | (.0532) | (.0515) |
| Size growth | $\hat{\beta}_{34}$ | -1.7152 | -2.3070 | 2.5060 | -6.0132 |
| | | (.4976) | (.5917) | (.6020) | (.4934) |
| Dummy durables | $\hat{\beta}_{41}$ | -.0230 | -.0392 | .0613 | .0728 |
| | | (.0334)* | (.0320)* | (.0228) | (.0180) |
| Dummy fashion | $\hat{\beta}_{42}$ | -.2331 | -.1957 | -.0023 | -.0059 |
| | | (.0329) | (.0333) | (.0240)* | (.0177)* |
| Time | $\hat{\beta}_{43}$ | -.0183 | -.0226 | .0020 | .0179 |
| | | (.0037) | (.0039) | (.0011)* | (.0010) |
| n. of observations | | 160 | 160 | 560 | 560 |
| Degree of fit | $\rho^2$ | .925 | .913 | .954 | .971 |

*Notes table 8.2:*
*- estimated standard errors are printed beneath the coefficients;*
*- an asterisk * is printed next to the standard error if for the coefficient* $\hat{\beta}$, $|\hat{\beta}| < 1.96\ \hat{\sigma}(\hat{\beta})$ *obtains, i.e. if the coefficient is not significantly different from zero at a 5% level of significance.*
*- the square of the correlation coefficient between the variable to be explained and its estimation* ($\rho^2$) *is used to measure the degree of fit.*

tax shields are hypothesized to be accompanied by the use of little debt. This hypothesis is supported for both debt categories in the Dutch data and for long term debt in the German data. For short term debt in the German data it is rejected.

Myers' growth effect and the signal effect of growth are tested with the average annual growth rate in size. Myers' growth effect is supported by the German data, in which a significantly positive coefficient of the growth variable is found in the analysis of short term debt and a significantly negative one for long term debt. The latter coefficient is also found in the Dutch data, but there the coefficient for short term debt is also negative. The signal effect, which is predominantly positive, is hardly supported by the data.

The hypothesis that firms selling durables use comparatively little debt in order to be able to meet possible future claims regarding service and maintenance of their products, i.e. Titman's durable effects, is empirically tested with the dummy variable for durables. In the Dutch data, negative coefficients are indeed found, but they are not significant. In the German data, significantly positive effects are found, contrary to the hypothesis, so that Titman's durable effect does not appear to be a major determinant of financial structure for the shoptypes investigated here. It should be noted, however, that Titman's argument refers to producers rather than to retailers, since future maintenance and repairs are usually guaranteed by the producer, not the retailer.

Finally, the time trend, for which no hypothesis is formulated, appears to have different effects, as was the case with assets. Both coefficients found in the Dutch data are significantly negative, while the analyses of German data produces positive time trends, albeit that one of them, i.e. the trend in short term debt, is insignificant. Again, there is no ready explanation for these results.

## 8.4 Analysis of changes in financial structure

### 8.4.1 *Hypotheses and variables*

In the previous section, financial structure is expressed in terms of money values, i.e. the amounts that appear in the firm's financial statement. Since many financing and investment decisions have effects that stretch well into the future, the amounts presented in a financial statement at any point in time can be regarded as resultants of many decisions in the past. Analysing financial structure expressed in these terms may, to a certain extent, obscure the behaviour which induces changes in financial structure, i.e. obscure the effects of single decisions or decisions in a short period of time. This is particularly true if the changes are small compared with the observed amounts, as is the case with the average data at hand. In order to expose the behaviour more sharply, this section comprises an analysis of the available data from a somewhat different perspective, i.e. first differences between the observed amounts are analysed. Taking first differences largely removes the effects of decisions made in the past that are still reflected in the present financial statement. For instance, changes in assets only measure new investments or depreciations and not the remaining value of assets that were purchased years ago. It should be noted, however, that the data only allow the calculation of year-to-year changes, which reflect the effect on balance of all changes in the period. Moreover, the data are shoptype averages and taking averages may conceal ("average out") the effects of decisions within firms; "average behaviour", if it exists at all, may be hard to explain. However, despite these limitations, an analysis of changes may be worth while, because it may throw some light on the intertemporal dynamics of small firm finance.

The simplest way to formulate hypotheses for the analysis of changes in debt categories would be to rewrite the previous section's hypotheses in terms of changes. In that case, most variables would be transformed into changes and the time invariant variables would be dropped from the list of hypotheses. Basically, that is the procedure followed here, but some adaptations are made. First of all, while most amounts in the firm's financial statement are inherently positive, changes in these amounts may be positive or negative. This should be taken into account because positive changes (e.g. investments) may have different effects than negative changes (e.g. depreciations). Secondly, not all variables have to be transformed into changes. Some variables may reflect the way in which a firm reacts to changes, i.e. reflect the firm's sensitivity for changes. The firm characteristics that describe this sensitivity are relevant as such, i.e. their values are relevant rather than the changes in their values. Hence, transformation of these variables is not called for. Thirdly, in the context of changes some variables may have a somewhat different purport than in the context of the previous analyses. Whenever necessary, these variables are redefined to reflect their different meaning.

The present analysis of changes is along the general lines sketched above, i.e. changes in debt categories are modelled so that they may spring from different sources and the degree in which these sources are tapped depends on certain firm characteristics. The sources of changes in debt can be grouped in three categories, viz. investments, repayments and rearrangements, while the sensitivity for these sources is expected to depend on the firm's asset structure, its inventory turnover, its cost structure and on market developments. Below, these elements are briefly elaborated.

Investments can be positive or negative (i.e. be investments or disinvestments) and they can refer to current assets and to fixed assets. In view of the support for the traditional rules of finance, that was found in the previous section, changes in current and fixed assets will probably not have the same effects on short and long term debt. Changes in current assets are likely to have a stronger effect on short term debt than on long term debt and the opposite is expected for changes in fixed assets. The probably different effects of investments and disinvestments are already mentioned, but it is hard to specify expectations as to the exact nature of the difference. Hence, the four possible changes in assets are separately included in the analysis, so that minimal restrictions are imposed on the outcomes.

Repayments of debt can take place from the funds generated by the firm's operations and from the proceeds of an equity issue. The latter cannot be included in the analysis, of course, because in the data at hand equity is the balancing item of debt and assets, so that the inclusion of equity would produce a definition relation. The funds generated by the firm's operations can be approximated by several variables (gross profits, EBIT, net profits), but the cash flow is probably the most widely used variable in this context. Since the level of debt is likely to be chosen in accordance with the capacity to generate cash flow, the change in cash flow is included in the present analysis of changes in debt. A growing cash flow will offer more repayment capacity, so a negative effect on debt is expected.

Rearrangements, the last source of changes in debt, refer to adjustments in the maturity structure of debt. Short term debt can be replaced by long term debt and vice versa. Rearrangements of debt categories can take place under the influence of changes in several variables, e.g. new securities can be given or the relative prices of debt categories may change. Unfortunately, the data comprise neither information on the term structure of interest rates nor on the securities involved. Consequently, rearrangements are accounted for by directly including in the analysis the changes in the complementary debt category, i.e. changes in long term debt are included as an explanatory variable in the analysis of changes in short term debt and vice versa. If both debt categories change independently, the coefficients of these variables will be zero and if rearrangements take place negative coefficients will appear.

The degree to which the above mentioned sources achieve changes in debt categories is likely to depend on certain firm characteristics. In essence, these characteristics should reflect the financial flexibility[9] of the firm or, in other words, its capacity to anticipate or react to changes in the financial environment. Flexibility is a qualitative aspect and, as most qualitative aspects, it is hard to capture in a numerical value. However, the three variables to be discussed below together may give a reasonable approximation of the concept of flexibility. The first proxy variable is the firm's <u>asset structure</u>, measured as the share of fixed assets in total assets. A large fixed asset share is attended with low flexibility, because funds tied up in fixed assets can be reallocated less frequently than investments in current assets. The second proxy variable, <u>inventory turnover</u>, has a similar effect as it indicates the liquidity of inventories which is, of course, positively related to flexibility. Hence, negative coefficients for asset structure and positive ones for inventory turnover would be expected. However, the traditional rules of finance broached earlier may complicate these simple effects, i.e. asset structure may have positive effects on changes in long term debt and inventory turnover may have negative effects on long term debt. The third and last variable approximating flexibility is the firm's <u>cost structure</u>, expressed as the share of fixed costs in total costs. Large commitments in terms of costs limit the use of funds for alternative purposes and, hence, the possibilities to react to changes, so negative coefficients are expected for this variable.

Finally, a variable, <u>sales growth</u>, is included to reflect market developments. In addition to the effects discussed in chapter 5, it can be noted that in growing markets, optimism tends to be the prevailing mood and risks are played down. Generally, such circumstances lead to a heavy use of debt, while in contracting markets opposite effects can be observed. Since data on e.g. consumption expenditures for the shoptypes at hand are not available for both countries, market developments are approximated by sales growth. This variable is calculated as the year-to-year percentage growth, not the average growth rate used is the previous section.

-------

[9] *cf. Diepenhorst* (1966) *and Wytzes* (1975) *who argued earlier that qualitative aspects as elasticity and flexibility should be taken into account in the evaluation of financial structure.*

### 8.4.2 *Specification and estimation results*

As is already suggested in the previous sub-section, the simplest way to specify a relation for the analysis of changes in debt categories would be to rewrite (8.1) in terms of first differences. In that case, most variables would be transformed into first differences, the time invariant variables would be dropped from the specification and the time trend would become a constant. Although that is basically the procedure followed here, some adaptations are made to incorporate the new hypotheses and to reflect the somewhat different purport of certain variables. The main adaptations are that the "size" variables are replaced by the "source" variables, and that some variables are transformed into changes. The variables to be explained (debt categories) and the source variables are thus transformed. These changes are defined as simple first differences, not percentage differences. Taking percentages would not be meaningful for one of the source variables, cash flow, which can be negative, and it would make the results more difficult to interpret since the bookkeeping symmetry would be lost. The following specification is used:

$$(8.2)\quad \Delta Y_{it} = \beta_1 + \left[ \beta_2 \Delta CFL_{it} + \sum_{j=1}^{4} \beta_{3j} \Delta A_{ijt} + \beta_4 \Delta Y^*_{it} \right] \left[ \prod_{l=1}^{3} X_{i\,lt-1}^{\beta_{5l}} \right] exp\ \beta_6 \Delta' Z_{it} + u_{it}$$

in which :

| | |
|---|---|
| $Y$ | =the variable to be explained (short and long term debt respectively) |
| $CFL$ | =cash flow (net profits plus depreciation charges), |
| $A$ | =asset category (investment and disinvestment in current and fixed assets, |
| $Y^*$ | =complementary debt category, |
| $X,Z$ | =explanatory variables described in the text, |
| $i$ | =index for shoptype, |
| $t$ | =index for time, |
| $\Delta$ | =first difference operator (e.g. $\Delta Y_t = Y_t - Y_{t-1}$), |
| $\Delta'$ | =percentage first difference operator (e.g. $\Delta' Y_t = (Y_t - Y_{t-1})/Y_{t-1}$), |
| $\beta$ | =coefficients to be estimated, |
| $u$ | =disturbance term. |

Comparing (8.2) and (8.1) we see that the variables to be explained and the size variable in (8.1) are transformed into changes and that the change in size is decomposed into four constituting elements (four "source" variables). Further, the changes in the complementary debt category and in cash flow are added to the linear part of (8.2). The dummy variables in (8.1) are, of course, not included in (8.2), as are the tax shields and size growth, whereas the time trend in (8.1) corresponds with the intercept in (8.2). The cost structure and sales growth are added compared with (8.1), while asset structure and inventory turnover have different purport in the context of changes than in the context of the previous section.

The technical details of (8.2) are essentially the same as those of (8.1). The variables specified as $X$ are divided by their average value in each year in order to remove changes in their average value over time. Sales growth, which can be negative, is specified as $Z$ in an exponent of the constant $e$. The disturbance term $u$ is again assumed to be a stochastic variable with an intertemporally independent distribution, zero expectation and constant variance. The correlation per shoptype over time, that was found in the previous subsection, is likely to be removed by the first difference operation[10]. A multiplicative specification is chosen for $X$ and $Z$ because such a specification allows for interaction between variables. Again, a non linear least squares fit based on Marquardt's (1963) algorithm is used to produce the estimation results, which are presented in table 8.3. Below, the results are briefly discussed.

The intercepts, which can be regarded as autonomous growth rates, appear to be of little importance. The coefficients do not show any pattern in their signs, and they are not significantly different from zero in three of the four cases. Repayment effects on changes in debt categories are indeed found: three coefficients of the changes in cash flow are significantly negative as hypothesized, the remaining one is insignificantly positive. The effects are much stronger in the Dutch data than in the German data. The coefficients obtained with the German data are, in fact, extremely small, both in absolute terms and in comparison with the Dutch results. The effects of investments are somewhat more complex than was anticipated. Changes in current assets, both investments and disinvestments, have stronger effects on short term debt than on long term debt, as was expected. With one exception, all coefficients referring to current assets are significantly positive. Investments in fixed assets appear to have no significant effects on short term debt and larger, significantly positive effects on long term debt, also conformably to the expectations. However, disinvestments in fixed assets have stronger effects on short term debt than on long term debt, contrary to the

-------

[10] *Inspection of the residuals of* (8.2), *printed in an* $I$ *times* $T-1$ *matrix, showed that this is indeed the case.*

expectations. Moreover, in the analyses of short term debt, disinvestments in fixed assets have opposite effects in both countries, while both coefficients are significantly different from zero. These results are hard to explain, particularly because the other effects of investments are largely as could be expected. The hypothesized rearrangements are only found in the Dutch data: changes in the complementary debt category have a significantly negative effect in both analyses. In the German data, the complementary debt category appears with positive coefficients, one of which is significant. Positive effects are not anticipated; they may point to a sort of "package deal" in debt financing, but this phenomenon is not mentioned in the literature as a common aspect of the supply of debt.

The variables included to reflect the firm's flexibility largely have the expected effects, although minor differences also occur. Asset structure and inventory turnover were expected to have negative and positive effects respectively, but the traditional rules of finance could reverse the effects on long term debt. The results obtained in the analysis of short term debt are in accordance with the hypotheses and significant in three of the four cases: the coefficients of asset structure are negative (although they are not significant in the German data) and those of inventory turnover are positive. In the analysis of long term debt, the German results reflect the traditional financing rule of synchronizing the terms of investment and finance: asset structure has a positive effect on changes in long term debt and inventory turnover has a negative effect. Both effects are significant. On the other hand, in the Dutch data no significant effects of these variables are found, while the signs of their coefficients are not in accordance with either of the two hypotheses. A similar remark applies to the effect of cost structure on long term debt in the Dutch data: the effect is insignificantly positive, contrary to the hypothesis. In the other three cases, cost structure has the expected negative effect, although the effect is insignificantly negative in the analysis of long term debt, German data. Finally, sales growth does not prove to have a particularly strong effect on changes in debt, as three of the four coefficients are not significantly different from zero. Only the analysis of long term debt, German data produces the expected positive effect.

The discussion of the estimation results is concluded with two remarks. First, when the results obtained with the Dutch and the German data are compared, it should be kept in mind that the German data refer to a longer period than the Dutch data, and that both periods are largely disjunct. Second, the year-to-year changes in long term debt may be small due to contracts which stretch over a longer period. In an analysis on the basis of a comparatively short time series, the variables describing these changes may have hardly appreciable effects. Those two factors could play a role in the rather frequent occurrence of insignificant coefficients in the analysis of long term debt in the Dutch data.

Table 8.3 Estimation results changes in short and long term debt

| | | Dutch data | | German data | |
|---|---|---|---|---|---|
| | | Δshort | Δlong | Δshort | Δlong |
| | | term debt | | term debt | |
| Intercept | $\hat{\beta}_1$ | 1.7703 | -.3905 | -3.3394 | 1.0366 |
| | | (2.5021)* | (1.9785)* | (.7790) | (.6383)* |
| ΔCash flow | $\hat{\beta}_2$ | -.4962 | -.4143 | .0039 | -.0068 |
| | | (.1562) | (.1000) | (.0031)* | (.0020) |
| Investments current assets | $\hat{\beta}_{31}$ | 1.0023 | .1634 | .3985 | .0423 |
| | | (.1723) | (.0674) | (.0283) | (.0195) |
| Disinvestments current assets | $\hat{\beta}_{32}$ | .3962 | .2614 | .4408 | .0357 |
| | | (.1576) | (.1068) | (.0418) | (.0312)* |
| Investments fixed assets | $\hat{\beta}_{33}$ | -.0508 | .4529 | -.0343 | .4737 |
| | | (.1004)* | (.0791) | (.0554)* | (.0408) |
| Disinvestments fixed assets | $\hat{\beta}_{34}$ | .7685 | .5725 | -.4676 | .3501 |
| | | (.2446) | (.1423) | (.0722) | (.0493) |
| Rearrangement (compl.debt cat.) | $\hat{\beta}_4$ | -.4179 | -.1946 | .1141 | .0549 |
| | | (.1190) | (.0563) | (.0535) | (.0305)* |
| Asset structure (fixed/tot.ass. ) | $\hat{\beta}_{51}$ | -3.5382 | -.7022 | -.1551 | .3922 |
| | | (.9320) | (.5403)* | (.1048)* | (.1265) |
| Inventory turnover | $\hat{\beta}_{52}$ | 1.0133 | -.1746 | .1748 | -.3134 |
| | | (.2684) | (.1337)* | (.0770) | (.1052) |
| Cost structure (fixed/tot.cost ) | $\hat{\beta}_{53}$ | -1.8032 | .4493 | -.3101 | -.2692 |
| | | (.8339) | (.7402)* | (.1469) | (.1796)* |
| Sales growth (%) | $\hat{\beta}_6$ | -.7577 | .5421 | .1442 | .3349 |
| | | (.7274)* | (.2816)* | (.0795)* | (.1129) |
| n. of observations | | 144 | 144 | 532 | 532 |
| Degree of fit | $\rho^2$ | .590 | .689 | .855 | .840 |

*Note table* 8.3: *see note table* 8.2

# PART IV
# applications and conclusions

# 9 APPLICATIONS

## 9.1 Introduction

Applications are not a major topic, to say the least, in studies that adhere to the theory of finance and if applications are included at all, their relation with the preceding theoretical analyses is generally a very loose one. As Tempelaar (1985) points out, many textbooks on finance start with rigorous analyses in the neoclassical framework, but the applications later on are usually confined to qualitative recommendations or checklists that have little to do with the foregoing arguments in the MM-tradition. The opposite can be observed in the practice of (small) business finance. The traditional financial statement analysis produced an enormous amount of rules of thumb, ratio pyramids, index systems, scoring models, discriminant models, etc., some of which are extensively used in practice but in which theoretical developments are virtually ignored. Attempts to bridge the gap between modern finance theories and traditional financial statement analysis have been made (e.g. Lev, 1974), but theory and practice still seem to be largely separate areas of inquiry.

From a theoretical point of view, applications are hampered by a number of factors. Theoretical concepts can be impossible to measure directly and the transition from these concepts to empirical variables is a large one. Issues of central concern to the theory, e.g. the existence of optimal capital structure, seem to play a much less important role in practice. Financing decisions are seldom made in isolation, but are usually combined with investment decisions. Many of the assumptions on which the theory rests are probably not met in practice (all sorts of "imperfections" may exist). All these factors bring about that theoretically derived valuation formulae cannot directly be applied in practice: prescriptions on how to maximize firm value by choosing a particular capital structure cannot be given. As a result, financial economists tend to shun applications[1] which, in turn, tends to confine the study of (optimal) capital structure to an issue of purely academic interest instead of making it a tool of management.

On the other hand, the practice of financial statement analysis is not very fast in adopting insights from the theory for similar reasons. Issues of

-------

[1] *Myers* (1984) *puts it even more strongly, stating that it seems presumptuous to advice firms on capital structure when we are so far from explaining actual decisions.*

central concern to financial analysts, such as liquidity or the flexibility offered by a particular financial structure, are simply assumed away in the theory. Practicians also find it difficult to recognize their observable variables in the concepts used in the theory. The complexity of financial decisions, which are always interwoven with operating decisions, are only dimly reflected in theoretical models. These factors prevent a free flow of ideas from theory to practice (and vice-versa). For instance, textbooks on financial statement analysis typically do not comprise an inventory of financial structure determinants (see e.g. Foster, 1978), although, in our opinion, that would be the first thing to expect is such a treatise.

An additional obstacle, that merits separate discussion, is the virtually ineradicable habit in financial statement analysis to express every quantity related to financial structure as a ratio. As is already mentioned in section 5.2, the main topics in financial statement analysis, including ratio analysis, are repeated over and over again in textbooks and, over many decades, they have been taught at business schools and courses. As a result, the concept of ratios is deeply rooted in financial statement analysis, which can be illustrated by quoting Horrigan (1965 p.568), who states that "it is inconceivable that accounting data can be analysed without transforming it into ratios, in one way or another". Of course, criticism of financial ratio analysis is almost as old as the technique itself[2], but in spite of severe criticism thinking in terms of ratios still pervades almost the entire area of financial statement analysis. This thinking in terms of concepts as simple and as restricted as a ratio blocks the adoption of all but the simplest of new developments and it prevents financial structure to be described in more than one dimension, i.e. as the net result of various leverage related costs and benefits.

From the above it will be clear that the application of research outcomes, such as those presented in the previous chapters, is a somewhat neglected subject in the literature. For instance, none of the 12 empirical studies mentioned in table 2.3 contains a section on applications[3]. Nevertheless, an illustration of the possibilities to apply the research outcomes is included in this study because we believe that the estimation

-------

[2]*To give but one example, Lev* (1974 *p.*36) *describes ratio analysis principles as "wholly inadequate. They are surprisingly naive, often ambiguous and usually unverified from an empirical point of view."*

[3]*Although scant, impulses to a different use of accounting data, based on theoretical considerations, can also be found in the literature; see Lev and Sunder* (1979) *for a good example of this.*

results discussed earlier are not only of scientific interest, but that they can also be used to improve the procedures of financial analysis and financial decision making that are currently used in practice. For instance, most financial analysts and decision makers in small business agree, that an industry average debt ratio is an inadequate measure to evaluate a firm's leverage, but still it is used in practice for lack of more precise measures. The theory of finance tells us that (optimal) leverage depends on e.g. collateral securities. Although this view is subscribed by most people in practice, collateral securities are seldom incorporated in an industry average standard, because empirical estimations of the influence of collaterals on leverage are not available and such estimations would not fit very well in the framework of ratio analysis. Our empirical analyses not only confirm the theoretical insight, they also provide a quantitative estimation of the influence of collateral securities on leverage in a sample. Clearly, this estimation does not describe the locus of optimal capital structure as a function of collateral securities, but the estimation can reasonably be expected to be closer to the optimum than the industry average debt ratio is. So, if a firm's leverage is to be evaluated, as is frequently done in practice, the logical measure to use, in our opinion, is an estimation based on all available determinants, instead of a simple industry average ratio.

In this way, theoretical developments and insights may find their way to practical applications. It should be noted, however, that any application is likely to be vulnerable to criticism from a theoretical point of view. Applying results means, so to speak, soiling ones hands. Clear concepts as optimal capital structure are traded in for vague notions as "closer to the optimum than...". Empirical estimations may include variables without a clear theoretical basis, the effects of various determinants may be mixed and in any application process a number of more or less arbitrary decisions have to be made.

As an example, the application of the research outcomes in interfirm comparisons is elaborated in the next sub-section. Of course, this example merely illustrates the possibilities, a ready to hand product is not provided. Such a product would require recent data and the development of software, which is beyond the scope of this study.

## 9.2 Modelling interfirm comparisons

The current practice

In small and medium-sized business, interfirm comparisons are a widely used instrument to assess the performance of individual firms and to provide the entrepreneur with information regarding various aspects of the firm's operations. In many countries of the European Community, information for interfirm comparisons is regularly published[4], and there are numerous private circles where such information is exchanged (e.g. within chain stores). Apparently, there is a large demand for these simple tools of management and entrepreneurs and small business consultants seem to base their judgements of the course of business to a considerable extent on this type of information.

The current practice of interfirm comparisons invariably involves ratio analysis in one way or another. This current practice, although diverse in form and contents, essentially consists of calculating ratios for an individual firm, comparing these with some industry-standard (usually industry average or median ratios) and drawing conclusions from the sign and the significance of the deviations. The conclusions generally are formulated in terms of "good or bad" or "favourable-unfavourable", with ample reservations made. Of course, it is also recognized in practice that certain firm characteristics or policy decisions affect the value of certain ratios. The influence of collaterals on leverage is already mentioned in the previous sub-section. Another generally recognized example is the firm's assortment policy, which can negatively influence inventory turnover if the policy is to carry a particularly wide or deep assortment.

In practice, this problem is dealt with in two ways. The first way is taking a firm's characteristics into account in the evaluation of the results of a ratio analysis. For instance, a significantly below average inventory turnover may not be considered an indication of bad performance if the firm carries a very wide or deep assortment. Clearly, this procedure is unsystematical and lacks the numerical basis that is needed for a sound interfirm comparison. The second way used in practice to untangle performance and characteristics is breaking down the industry average by firm characteristics, e.g. using ratios per size class instead of one overall ratio. This method has considerable disadvantages too. When industry averages are broken down by a continuous characteristic (e.g. size), arbitrary classes

-------

[4] *For instance, the EIM has been has been publishing data for interfirm comparisons for almost half a century; another example are the German data used in chapter* 7.

must be formed, meaning the loss of information. It also means that the industry average ratio jumps from one value to another if the boundary-line between two classes is crossed, while the influence of a continuous characteristic is more likely to be gradual. Moreover, a breakdown by one characteristic almost inevitably also implies a breakdown by other characteristics. For instance, in most retailing industries assortment composition varies with size, so different size classes are likely to have a different assortment composition too. If this is the case, it remains unclear whether differences in e.g. inventory turnover are caused by size differences or by assortment differences between the classes. Moreover, it means that a firm may belong to one size class as regards size but to another size class as regards assortment composition so that it is unclear which ratio should be used. The latter problem can be overcome by breaking down the industry average by a number of characteristics, e.g. assortment classes per size class. However, this procedure can become quite cumbersome and data consuming. For three characteristics, each divided in five classes, the total number of classes is $5^3=125$, so the required number of observations is a multiple of 125.

Discussion

In addition to the practical objections discussed above, the efficacy of the current practice can be questioned from two more fundamental points of view. The first and most fundamental question relates to all forms of interfirm comparison and concerns the usefulness of standards based on information from other firms. As is already discussed in the previous sub-section, these standards clearly cannot be considered optimal values and, hence, deviations from them cannot be interpreted, strictly speaking, in terms of "good or bad". However, in addition to the arguments provided in the previous sub-section, it can be noted that these standards may also, and more fruitfully, be looked upon in several other ways than optimal or target values. Industry standards provide, without requiring prior information, an external "best guess" of the firm's costs, profits, financial structure, etc. This information can be used to determine on which aspects the firm deviates -by choice or not- from the industry, i.e. depict the firm's relative position in the industry. It can also be used as an approximation of information that is hard or expensive to obtain in other ways (e.g. through market research). This is especially relevant for the effects of changes in firm policy, for which no internal data are available (e.g. including a new product group in a shop's assortment). Instead of collecting and processing data for a decision, or arriving at a decision through trial and error, one can, so to speak, look directly at the result of these decision making processes in other firms. In this way, interfirm comparisons can be used to assess the firm's relative performance and, subsequently, to formulate targets for future performance. This is particularly valuable in small business, because small firms usually lack the capacity to obtain such information in other ways. It is also particularly possible in small business because there are large numbers of similar small firms in almost all industries.

If interfirm comparison is considered useful, the second question is which standards are adequate for assessing the performance of individual firms. After the discussion in the previous sub-section it will obvious that ratios are not adequate. Ratios imply a well defined relation between the items in their numerator and denominator, and this relation is restricted in both form and contents. It is restricted in form because ratios represent a fixed proportion, equivalent to a linear homogeneous relation, which does not allow for scale-effects. It is restricted in contents because ratios comprise only two variables, while usually it is known in advance that more variables play a role in the phenomenon the ratio seeks to express. For instance, as we have seen in the previous chapters, the amount of debt does not only depend on size (total assets), but also on asset composition, collateral security, etc. Similarly, labour volume does not only depend on sales size, but also on wage rate assortment composition, etc. In ratio-analysis, all these dependencies are either ignored or inadequately dealt with by e.g. forming size-classes within an industry, or by involving additional variables in the evaluation of the results of a ratio-analysis

Modelling interfirm comparisons

The limitations of the usual, ratio-based methods of interfirm comparison can be overcome by using the more general relations developed in the previous chapters to construct industry standards. In these relations, the influence of various firm characteristics on financial structure is modelled, so these relations can be used to provide a "best guess" of the financial structure of a firm with a specific combination of characteristics. Since each individual firm is likely to have its own combination of these characteristics, the models allow separate industry standard debt ratios to be calculated for each individual firm. Similar models are specified and estimated for all major items of the income statement and balance sheet, in order to provide a complete picture. To illustrate, this procedure is elaborated for one of the samples analysed in chapter 7, viz. the sample of stationer's shops. Although, in principle, the estimation results obtained in chapter 7 could be used, additional estimations are necessary to include items of the income statement, so a new estimation round is performed. The following specification is used:

$$(9.1)\quad Y_{ik} = \beta_1 + \sum_{h \in R_k} \beta_{2h} S_{hik} \left[ \prod_{j \in V_k} X_{ij}^{\beta_{3j}} \right] exp \sum_{l \in W_k} \beta_{4l} Z_{il} + u_i$$

in which :

$Y$ =the variable to be explained (described below),

$S$ =size variable (described below),

$X,Z$ =explanatory variables described in the appendix to this chapter,

$R_k$ =the set of indices for size variables $S_h$ used in the explanation of variable $Y_k$ (sales in 5 groups for current assets and all items of the income statement, sales and sales x dummy own buildings for fixed assets and total assets for debt categories),

$V_k$ =the set of indices for the explanatory variables $X_j$ used in the explanation of variable $Y_k$,

$W_k$ =the set of indices for the explanatory variables $Z_l$ used in the explanation of variable $Y_k$,

$i$ =index for firm,

$k$ =index for the variable to be explained,

$\beta$ =coefficients to be estimated,

$u$ =disturbance term.

Since (9.1) is the same as (7.1), this specification needs no further discussion. Recall, however, that in extreme simplification (9.1) is equivalent to a ratio. For instance, if $Y_k$ is short term debt, then $\beta_2$ is equivalent to the ratio of short term debt to total assets if all other coefficients in the equation do not differ significantly from zero. If scale effects appear ($\beta_1$ significantly different from zero), $\beta_2$ can be interpreted as the scale-adjusted ratio of short term debt to total assets. For the full equation, $\beta_2$ represents the scale-adjusted ratio for firms with average values of $X_i$ and zero values for $Z_j$. This equivalence to ratios is a major advantage when a complex relation as (9.1) is introduced in practice, where ratio analysis is virtually the only technique used. It offers the possibility to start with a familiar ratio and, by generalizing the formula and subsequently adding variables, to "build up" (9.1) in a step-by-step way. Experience with this procedure on various occasions has been positive.

Formula (9.1) is estimated for the following variables:

| Income statement | Balance sheet |
|---|---|
| –gross profit | –inventories |
| –labour volume | –remaining current assets |
| –floorspace | –fixed assets |
| –remaining costs | –short term debt |
| | –long term debt |

Note that the choice of these variables is rather arbitrary: total costs can be split up in more categories and more separate items on the balance sheet can be distinguished. However, for the purpose of this study the variables mentioned above provide a detailed enough picture. The relations for labour volume and floorspace are adopted from Nooteboom (1980) and Thurik (1984), who analyse retail productivity in great detail. Note that multiplication of labour volume by wage rate and by per cent paid labour produces labour costs, and that multiplication of floorspace by housing costs per square metre produces housing costs. Together with remaining costs, these cost categories constitute total costs.

The estimation results are presented in tables 9.1A and 9.2A in the appendix to this chapter. Combined with table 9.3A, these tables also show the specification of (9.1) in terms of variables. Compared with the analyses in the previous chapter, insignificant variables are dropped from (9.1), and the assortment composition and collateral securities are included in a more detailed form. A more comprehensive discussion of similar models can be found in the literature (Nooteboom (1980), Thurik (1984), Thurik and Van der Wijst (1984), Van der Wijst (1985), Amiabel et al. (1986)). Together, the models specified in the appendix enable a fairly detailed and complete assessment of the performance of a small firm by comparing the actual data of the firm with the industry standard values calculated for this specific firm. The results of this assessment may differ from the outcomes of a traditional ratio-analysis, as is shown below.

## An application

The models described above can be used for interfirm comparisons by simply substituting the relevant data for a particular firm into the equations and calculating the result. This results represents the industry standard for this particular firm. To illustrate , an example from the 116 stationer's shops analysed in this chapter is briefly elaborated. The relevant data of this example firm and the corresponding industry average are presented in table 9.3A in the appendix to this chapter. For the sake of simplicity, it is assumed that all dummy variables, except the one for recent investments, equal zero. Table 9.3A shows that the example firm is somewhat larger than the average and that it is specialized in the professional segment of the market (particularly office furniture and machinery). This specialization affects the wage rate (and, hence, labour productivity) and the housing costs per $m^2$ (and, hence, floorspace productivity), because the professional market requires better qualified employees and a showroom to demonstrate the office equipment. Also, in the professional market it is more customary than in the consumer market to sell for future delivery (i.e. less sales are made from stock) and more customary to sell on account; this, of course, affects current assets. In addition to this, the example firm does not own its buildings and it invested comparatively little in recent years, which affects fixed assets and debt. The latter is also influenced by lack of collateral security.

To see how all these considerations are reflected in the estimation results, consider e.g. current assets. From table 9.3A it can be calculated that the industry average ratio of inventories-to-sales is .167 and that the ratio of remaining current assets-to-sales is .136. In the analyses, these ratios are "split up" by assortment groups, as table 9.1A shows (the intercept and dummy deep assortment are ignored here). Since office furniture and machinery are usually sold for future delivery and on account, the coefficient of this assortment group is much lower than the average ratio for inventories, but much higher in the case of remaining current assets. Hence, if the average ratios are used for a firm specialized in the professional segment of the market, a misleading picture may be obtained, as becomes clear below.

All the above considerations are taken into account when the models are applied to the example firm. In table 9.1, the results of the models are confronted with the data from the example firm on the one hand and with the results of a traditional ratio-analysis on the other hand. (For the latter, the industry average ratios based on total sales and total assets are used, which can be calculated from table 9.3A.) Table 9.1 shows that, taking the above mentioned dependencies into account, it can be concluded that the example firm's gross margin is comparatively low (some 2 percentage-points below its industry standard), as are the labour costs and the remaining costs. Ignoring the above considerations regarding the income statement, by using ratio-analysis, it would be concluded that the example firm has comparatively low housing costs and high remaining costs and that gross profits and labour costs are in line with the industry standard (ratio). So, in the case of the remaining costs the conclusion is reversed.

Table 9.1 Model results versus data from the example firm and the results of a ratio–analysis

| description | data example firm | model–results | results ratio–analysis |
|---|---|---|---|
| | in Guilders | | |
| <u>income statement</u> | | | |
| gross profit | 466,000 | 493,775 | 467,600 |
| labour costs | 147,162 | 171,754 | 143,220 |
| housing costs | 57,000 | 56,583 | 80,640 |
| remaining costs | 117,200 | 131,009 | 107,198 |
| <u>balance sheet</u> | | | |
| inventories | 206,000 | 178,700 | 233,800 |
| remaining current assets | 206,500 | 282,600 | 190,400 |
| fixed assets | 147,500 | 187,500 | 308,000 |
| short term debt | 142,500 | 159,500 | 188,200 |
| long term debt | 26,500 | 37,300 | 91,800 |

Regarding the balance sheet, the models show that the example firm has higher inventories but lower remaining current assets than could be expected on the basis of the firm's characteristics (specialization). Using ratio–analysis the conclusion would again be reversed (low inventories and high remaining current assets). The reason is, of course, that ratio–analysis ignores the fact that the example firm realizes a far above average part of its sales in office furniture and machinery, which require little inventories (since these products are not sold from stock), but a large amount of remaining current assets (since these products are sold on account). Regarding the fixed assets and both debt categories, the models and ratio–analysis point in the same direction, but the former technique results in far less pronounced differences with the actual data than the latter.

Obviously, the action taken by the entrepreneur (or the advice of a consultant) will strongly depend on the type of analysis performed. The more comprehensive analysis provided by the models suggests that attention should be centred on the gross margin and the inventories. If, on the other hand, ratio–analysis is used, attention is likely to focus on the remaining cost and the remaining current assets which are, in fact, already below the industry standard.

Conclusion

In the foregoing, it is demonstrated how the research outcomes, obtained in the previous chapter, can be applied to the widely used practice of interfirm comparisons. In our opinion, this application produces an improvement upon the current practice of interfirm comparisons. The calculations involved in the extended procedure described here are somewhat more complex, but since most data for interfirm comparisons are stored in computers anyway, this is hardly an objection. To the user of a computerized system, the complexity of the calculations is to a large extend irrelevant.

As regards the actual application of the procedure described here, it can be reported that steps are being taken by the EIM to implement the procedure in its system of interfirm comparison. This requires extensive preparations since more data per firm have to be collected than is presently the case. After a successful pilot investigation, application in a chain store is under consideration. In several industries, preliminary applications are being prepared. Also, after a pilot study reported on by Broeksteeg et al. (1987), the first steps to implement the procedure in the loan evaluation process of a bank have been taken.

Table 9.1A Estimated coefficients of (9.1), stationer's shops

| | | gross profits | labour volume (hours) | floor-space ($m^2$) | remain-ing costs | invent-ories | remain-ing current assets |
|---|---|---|---|---|---|---|---|
| intercept | $\hat{\beta}_1$ | – | 1073 (533) | 21 (22)* | 14389 (4360) | 24159 (15902)* | −2332 (15411)* |
| S1 | $\hat{\beta}_{21}$ | .758 (.045) | 22.9 (3.0) | .656 (.106) | .158 (.023) | .186 (.077) | .391 (.079) |
| S2 | $\hat{\beta}_{22}$ | .329 (.017) | 11.8 (1.1) | .204 (.040) | .043 (.009) | .146 (.029) | .144 (.031) |
| S3 | $\hat{\beta}_{23}$ | .439 (.023) | 14.6 (1.7) | .431 (.073) | .042 (.014) | .125 (.046) | .114 (.049) |
| S4 | $\hat{\beta}_{24}$ | .309 (.019) | 4.9 (1.1) | .177 (.036) | .091 (.009) | .080 (.028) | .202 (.030) |
| S5 | $\hat{\beta}_{25}$ | .234 (.012) | 6.7 (.08) | .150 (.036) | .058 (.007) | .141 (.025) | .070 (.026) |
| relative sales size | $\hat{\beta}_{31}$ | −.053 (.025) | – | – | – | – | – |
| wage rate | $\hat{\beta}_{32}$ | – | −.54 (.122) | – | – | – | – |
| housing costs per $m^2$ | $\hat{\beta}_{33}$ | – | – | −.801 (.090) | – | – | – |
| dummy deep assortment | $\hat{\beta}_{41}$ | – | – | – | – | .202 (.086) | – |
| degree of fit | $\rho^2$ | .961 | .827 | .808 | .824 | .547 | .736 |

*Notes table 9.1A:*
- *the estimation results are produced by a non-linear least squares fit using Marquardt's algorithm. See Marquardt* (1963);
- *estimated standard errors are printed beneath the estimated coefficients;*
- *an asterisk * is printed next to the standard error if for the coefficient $\hat{\beta}$, $|\hat{\beta}| < 1.64\ \hat{\sigma}(\hat{\beta})$ obtains, i.e. if the coefficient is not significantly different from zero at a* 10% *level of significance.*
- *the square of the correlation coefficient between the vector of $Y_i$ and its estimation is taken as a measure of the degree of fit*
- *estimation results for fixed assets are not included in the table these results are, after taking the proper dummies into account: fixed assets = −25+.1518 x sales. (see Van der Wijst,* 1987 *for more details)*

Table 9.2A Estimated coefficients of (9.1), stationer's shops

| description | | short term debt | long term debt |
|---|---|---|---|
| Intercept | $\hat{\beta}_1$ | 23.1 | 17.0 |
| | | (14.8)* | (10.6)* |
| Total assets | $\hat{\beta}_2$ | .196 | .056 |
| | | (.031) | (.019) |
| Assets structure | $\hat{\beta}_3$ | -.366 | .975 |
| | | (.056) | (.319) |
| Recent investments | $\hat{\beta}_{41}$ | .082 | .037 |
| | | (.015) | (.017) |
| Dummy mortgages | $\hat{\beta}_{42}$ | – | .730 |
| | | | (.246) |
| Dummy other securities | $\hat{\beta}_{43}$ | .330 | .519 |
| | | (.091) | (.142) |
| Dummy fiduciary transfer of property | $\hat{\beta}_{44}$ | .297 | – |
| | | (.090) | |
| degree of fit | $\rho^2$ | .799 | .645 |

*Note table 9.2A: see notes table 9.1A.*

Table 9.3A Data of the example firm and industry averages

| description | | example–firm | industry average |
|---|---|---|---|
| assortment composition: | | | |
| –printing and repairs | S1 | 10% | 5% |
| –professional office equipment | S2 | 30% | 25% |
| –stationery | S3 | – | 20% |
| –office furniture and machinery | S4 | 60% | 18% |
| –books, magazines, toys, etc. | S5 | – | 32% |
| | | | |
| gross margin | | 33,3% | 33,4% |
| employees (full–time equivalents) | | 5.25 | 5.60 |
| floorspace | | 600 $m^2$ | 289 $m^2$ |
| percentage paid labour | | 71 | 60 |
| total sales ($f$ $10^3$) | | 1400,0 | 965,0 |
| remaining costs ($f$ $10^3$) | | 117,2 | 73,2 |
| wage rate ($f$ per hour) | | 21,0 | 15,5 |
| housing costs ($f$ per $m^2$) | | 95,0 | 192,0 |
| recent investments ($f$ $10^3$) | | 40,0 | 72,4 |
| inventories ($f$ $10^3$) | | 206,0 | 160,0 |
| remaining current assets ($f$ $10^3$) | | 206,5 | 130,0 |
| fixed assets ($f$ $10^3$) | | 147,5 | 210,0 |
| long term debt ($f$ $10^3$) | | 26,5 | 82,0 |
| short term debt ($f$ $10^3$) | | 142,5 | 168,0 |

This study investigates the small firm's choice of the level and composition of its investments and of the sources from which the funds for these investments are attracted, or in short: the small firm's financial structure. This subject is studied from several points of view. The literature is inventoried to collect the suggestions that have been made regarding the determinants of financial structure. These suggestions are tested, as far as possible, in empirical analyses and the applicability of the results of these analyses is investigated.

Before addressing these issues, this study starts with a discussion of the characteristics of small business, specifically the financial characteristics, in chapter 2. Small business appears to be an ambiguous concept as it is defined differently in different countries and industries. The upper limit of what is considered to be small or medium-sized varies from 20 to 500 employees. Irrespective of the precise definition used, however, the small business sector is an important part of the economy in all western countries in terms of number of firms, employment, contribution to the national income, etc. In terms of finance and investment, the small business sector in The Netherlands appears from some tentative calculations to be at least as large as the domestic part of the firms listed at the stock exchange, but it is probably larger.

Not many specific characteristics of small business finance are consistently mentioned in the literature. Empirical studies of financial differences between small and large firms show no pattern in the size effects on financial structure; a small size is reported to be associated with high as well as low as well as average debt ratios. The discussion of the so-called small firm effect in stock pricing models produced many valuable suggestions as to the possible causes of this effect, but no satisfactory explanation of the effect itself. A description of the small firm's capital markets did not reveal many differences between small and large business finance either, although small firms were seen to raise their money in comparatively intransparent, small scale capital markets on which informal investors, as friends and relatives, play an important role. On the basis of this it is concluded, that apparently there are not that many differences between small and large business finance. Although specific markets and financial instruments may be limited to a certain size range, the combined effects of these limitations appear to cancel out, on balance. As the size of transactions decreases, various marketability and cost considerations seem to induce professional investors to gradually withdraw from the small firm's capital markets, but, enabled by the small amounts involved, informal investors successively take their place.

In the second part of this study, an inventory of financial structure determinants, collected from both the theoretical and the practical literature, is presented. In the theoretical literature, some broad categories of debt ratio determinants are distinguished, which describe the influence on financial structure of :

- taxes (corporate and/or personal),
- bankruptcy costs,
- agency costs (associated with debt, equity, customers, etc.) and
- the signalling effect.

These determinants are discussed in chapter 3, along with their empirical implications and the variables that are commonly used to approximate these determinants in empirical analyses.

Subsequently, theoretical aspects of the debt financing decision are studied in chapter 4 by constructing some models of capital structure choice. In these models, major recent developments in the theory of optimal capital structure are systematically restated in a mathematical form, based on a uniform framework of definitions and derivations. Starting with Modigliani and Miller's seminal work, the models are developed in a step-by-step way. Each subsequent step highlights a different aspect of the debt financing decision by relaxing one or more assumptions of the MM-proposition and analysing the effects on the valuation of corporate claims. In this way, the effects of:

- limited liability and the risk of default,
- bankruptcy costs,
- agency costs and
- informational differences

are discussed and modelled in a single-period setting as well as in a multi-period setting.

The theoretical exercises give rise to rather skeptical conclusions, because their most striking aspect is the large number of models which produce indeterminate results: there is no question of a tree-structured family of interrelated models. In fact, of the five paths investigated, only one path is practicable to the end, viz. the line of single-period models in which bankruptcy costs are specified as a function of the cash flow. The other four paths peter out in indeterminate results, often in an early stage of the analysis. So it can be concluded that, even within the simple and rather restrictive framework of assumptions adopted in this study, the robustness of the various theoretical models leaves much to be desired. The models appear to be sensitive for small (and by themselves trivial) changes in the specification of variables so that the prospects of extending the models with more realistic assumptions are questionable, to say the least. These aspects are presented with some emphasis, because they are hardly reported on in the literature.

However, models with determinate results are also obtained, in which various aspects of the debt financing decision are incorporated and in which the firm's optimal leverage (optimal capital structure) and maximal leverage (corporate debt capacity) are seen to be:

- positively influenced by the tax rate;
- negatively influenced by the bankruptcy costs (implicitly or explicitly specified) and the agency costs of debt;
- ambiguously influenced by the risk free rate of return and the characteristics of the distribution of the cash flow.

In chapter 5, the financial structure determinants from the practice of (small) business finance are discussed. The subjects addressed and the methods used in the practical literature vary widely but the following categories can be distinguished:

1. influences pertaining to the individual firm:
   –characteristics of the firm itself;
   –characteristics of the entrepreneur;
2. influences pertaining to the relationships among firms;
3. influences pertaining to trade and industry in general.

In each of these categories, a large number of influences on the firm's financial structure, collected from many sources, is discussed.

The similarities and differences between the determinants of financial structure presented in both the theoretical and practical literature are discussed in chapter 6. Comparing the determinants from the two sources, it can be concluded that both approaches differ more in method than in content. As regards content, theory and practice are largely concerned with analogous phenomena, but as regards method there are marked differences. The theoretical approach applies strict rules to derive conclusions from assumptions, whereas the practical literature arrives at conclusions using many different methods such as reasoning, empirical facts and research outcomes. As a consequence, the nature of the conclusions differs rather strongly between both approaches. In the theoretical approach, simple and straightforward conclusions are derived on the basis of a large set of complex assumptions. By contrast, in the practical literature complex, often entangled and not always flawless conclusions are formulated on the basis of simple, if any, explicit assumptions (and, occasionally, complex implicit assumptions). The overall conclusion is that the theoretical and practical approach supplement each other, both in describing the financial behaviour of (small) firms and in providing a basis for empirical analyses.

In the third part of this study, two series of empirical analyses are performed. In chapter 7, data of individual firms within more or less narrowly defined industries are analysed. The 14 samples used in this study mainly refer to the retailing trade (11 samples), but samples from the metal products industry, the printing industry and structural engineering are also included. Average sales size of the samples varies from ƒ169.000,- to ƒ5.3 million and average total assets ranges between ƒ105.000,- and ƒ3.6 million. Because the samples contain detailed information of the firms involved, 15 different hypotheses can be tested, albeit in different subsets in different samples. Separate analyses for current and fixed assets and for short and long term debt are carried out in a specification that, on the one hand, allows adequate expression of the determinants from the literature and that, on the other hand, closely resembles commonly used financial ratios.

In the analyses of assets, important determinants of current assets appear to be size (scale effects), assortment composition and interest rate, while service (gross margin), substitution effects and assortment policy are less important. Fixed assets primarily depend on the ownership of buildings, the substitution of capital for labour and service (gross margin), while scale-effects and housing costs per square metre have a weaker influence.

Important determinants of long and short term debt (besides, of course, size) appear to be asset structure and inventory turnover (although these variables reflect traditional financing rules rather than theoretical considerations) and the dummies reflecting collateral securities (mortgages and other collaterals). To a lesser degree, the entrepreneur's age, recent investments, investment plans and family loans are also relevant. Scale effects, tax shelter effects, life cycle effects and the effect of limited liability appear to be of little importance, while the hypotheses regarding growth and co-ownership are mainly rejected. The (scant) empirical evidence on Myers' pecking order effect is mixed.

Comparing the results obtained in the different samples, it can be observed that the two industries which have the lowest asset requirements (retail trade in paint and shoemakers) also show the lowest degree of fit in the analyses of short term debt. Apparently, the very small amounts of short term debt in these samples are hardly subject to the primarily economic considerations tested here. The degree of fit of the analyses of long term debt in these samples is high and a similar remark applies to the analyses in this chapter in general.

In chapter 8, a Dutch and a German database of industry averages in the retail trade are analysed. Both databases contain time series data which, for this analysis, are pooled per country and the combined cross section and time series data are analysed in two ways. First, financial structure (in terms of money amounts representing the value of asset and debt categories) is analysed. Again, separate analyses for current and fixed assets and for long and short debt are carried out in a specification similar to the one used in chapter 7. Second, first differences in financial structure are analysed in order to expose more sharply the behaviour which induces the year-to-year changes in firm finance. The analysis of first differences is confined to debt categories.

In the analysis of financial structure, important variables for the explanation of the asset categories (current and fixed assets) prove to be the type of products sold (foods and perishable goods versus other products), the "production process" in terms of the employed combination of labour and capital and, of course, the size variable (sales). The price level of housing costs and the service level (approximated by the gross margin) also appear to be relevant. In the analysis of long and short term debt, the variables representing the collateral value of assets prove to be important determinants of financial structure. Asset structure (% fixed assets in total assets) is positively related to long term debt and negatively to short term debt, while opposite results are obtained with the second proxy variable for the collateral value of assets, inventory turnover. Thus, these variables influence the maturity structure of debt rather than the overall debt ratio. The third proxy variable for collateral security, a dummy variable for shoptypes selling products subject to fashion, appears to be relevant only in the Dutch data. The hypothesis that firms which sell durables use comparatively little debt, i.e. Titman's durable effect, is not supported in the Dutch data and rejected in the German data. Myers' growth effect is supported in the German data and partly supported in the Dutch data. The time trend and the tax shields produce results that differ in sign and significance.

The analysis of first differences in financial structure is modelled so that changes in debt categories may spring from different sources and that the degree in which these sources are tapped depends on certain firm characteristics. The sources of change are grouped in three categories, viz. investments, repayments and rearrangements, while the sensitivity for those sources is expected to depend on the firm's asset structure, inventory turnover and cost structure and also on market developments. The effects of investments are somewhat more complex than was anticipated, particularly as regards disinvestments in fixed assets. Repayment and rearrangement effects are virtually limited to the Dutch data. The firm characteristics representing the sensitivity for changes appear to be more relevant for short than for long term debt and also more relevant for the German data than for the Dutch data.

Throughout the analyses, scale effects are found to be of minor importance. A general finding is also that the degree of fit of the analyses is high, so that the approach followed in this chapter can be concluded to explain a large part of the differences in financial structure and in the changes (first differences) in financial structure between the shoptypes investigated here.

Finally, the possibilities to apply the results, that are obtained in the preceding chapters, are explored in chapter 9. It appears that applications are a somewhat neglected subject in the literature and obstacles, on both the theoretical and the practical side, are discussed. The possibilities to apply the research outcomes to the widely used practice of interfirm comparisons are illustrated, using the sample of stationer's shops as an example. It is shown that, in some instances, this application leads to conclusions that are the diametrical opposite of the conclusions produced by the traditional, ratio-based methods of interfirm comparison. On the basis of this, it is concluded that this application leads to an improvement upon the current practice of interfirm comparisons, and the steps, that are being taken to introduce this improved procedure into existing interfirm comparison systems, are described.

## REFERENCES


Adams, A., 1982, Barriers to product innovation in small firms: policy implications, European Small Business Journal, 1, 1, 67–86

Amiabel, J.B., A.J. van der Velden en D. van der Wijst, 1986, Bedrijfsdoorlichting detailhandel, (COB/SER, Den Haag).

Bannock, G., 1986, The economic role of the small firm in contemporary industrial society, in: James Curran, John Stanworth and David Watkins (ed.), The survival of the small firm 1, the economics of survival and entrepreneurship, (Gower, London), 8–18.

Barnea, A., R.A. Haugen and W. Senbet, 1981, An equilibrium analysis of debt financing under costly tax arbitrage and agency problems, Journal of Finance, vol. XXXVI, no. 3, June, 569–581.

Barry, C.B. and S.J. Brown, 1984, Differential information and the small firm effect, Journal of Financial Economics, vol. 13, 283–294.

Baxter, N.D., 1967, Leverage, risk of ruin and the cost of capital, Journal of Finance, September, 395–403.

Baxter, N.D. and J.G. Cragg, 1970, Corporate choice among long–term financing instruments, The Review of Economics and Statistics, vol. LII, no. 3, August, 225–235.

Bergh, W.M. van den, P.H.A.M. Verhaegen en R.E. Wessels, 1985, Vermogensstructuur in economisch perspectief, in: W.M. van der Bergh, P.H.A.M. Verhaegen en R.E. Wessels (red.), Financiering en belegging, stand van zaken anno 1985, (Erasmus Universiteit, Rotterdam).

Bergh, W.M. van den, J.W. Boogaart en A.I. Diepenhorst, 1984, Het grootte–effect in Nederlandse aandelenrendementen, in: W.M. van den Bergh, J. van der Meulen, G.J. Ruizendaal en P.H.A.M. Verhaegen (red.), Financiering en belegging, stand van zaken anno 1984, (Erasmus Universiteit, Rotterdam), 111–129.

Binks, M., and A. Jennings, 1986, Small firms as a source of economic rejuvenation, in: James Curran, John Stanworth and David Watkins (ed.), The survival of the small firm 1, the economics of survival and entrepreneurship, (Gower, London), 19–38.

Birch, D., 1981, The job generation process, Public Interest, vol. 65, 3–14.

Black, F. and M. Scholes, 1973, The pricing of options and corporate liabilities, Journal of Political Economy, 81, 637–659.

Braaksma, R.M., H.J. Gianotten, en W.H.J. Verhoeven, 1987, Midden– en Kleinbedrijf in de markt, een sterkte–/zwakteanalyse van de produktiestructuur, (Economisch Instituut voor het Midden– en Kleinbedrijf, Zoetermeer).

Bradley, M., G.A. Jarrell and E.H. Kim, 1984, On the existence of an optimal capital structure: theory and evidence, Journal of Finance, vol. XXXIX, no. 3, July, 857–880.

Brigham, E.F., and K.V. Smith, 1967, Cost of capital to the small firm, The Engineering Economist, vol. 13, no. 1, fall, 1–26.

Brigham, E.F., 1967, The cost of capital to the small firm, in: I. Pfeffer (ed.), The financing of small business: a current assessment (The MacMillan Company, New York), 97–125.

Broeksteeg, A.G.M., C.A. van Iperenburg en D. van der Wijst, 1987, Bedrijfsdoorlichting in de grafische industrie, Economisch Statistische Berichten, jrg. 72, no. 3619, 19 augustus, 765–770.

Castanias, R., 1983, Bankruptcy risk and optimal capital structure, Journal of Finance, vol. XXXVIII, no. 5, December, 1617–1635.

Chen, A.H., and E.H. Kim, 1979, Theories of corporate debt policy: a synthesis, Journal of Finance, vol. XXXIV, no. 2, May, 371–387.

Churchill, N.C., and V.L. Lewis, 1983, The five stages of small business growth, Harvard Business Review, May–June, 30–50.

Claessens, D., 1982, Bedrijfsbeheer in KMO en management in grote ondernemingen, Working paper, Katholieke Universiteit Leuven.

Copeland, T.E., and J.F. Weston, 1983, Financial theory and corporate policy, (Addison-Wesley, Reading(MA)).

Dailey, R.C., T.E. Reuschling and R.F. DeMong, 1977, The family owned business: capital funding, American Journal of Small Business, vol. II, no. 2, October, 30-39.

DeAngelo, H., and R. Masulis, 1980, Optimal capital structure under corporate and personal taxation, Journal of Financial Economics, March, 3-29.

Diepenhorst, A.I., en H. Willems, 1966, De optimale financiele structuur, in: C.F. Scheffer (red.), Kernproblemen der bedrijfseconomie, (Agon Elsevier, Amsterdam).

Donaldson, G., 1961, Corporate debt capacity, Graduate School of Business Administration ( Division of Research ) (Harvard University, Boston (MA)).

Donaldson, G., 1969, Strategy for financial mobility, Graduate School of Business Administration ( Division of Research ) (Harvard University, Boston (MA)).

Donckels, R., J. Degadt, P. Michel en L. Bragard, 1987, Financieel beleid en financiering van KMO's in Belgie, (W. Brondeel, Leuven).

Dorsman, A.B., J. van der Hilst and H.P.A.J. Lagendijk, 1987, Small firm effect and accounting data, in: W.G. Hallerbach, F.C.J. Bemelman, J.W.R. Schuit and P.J. Van Mierlo (ed.), Finance and investment, state of the art 1987, volume 10, (Erasmus University, Rotterdam), 173-189.

Eijgenhuijsen, H.G., J. Koelewijn en H. Visser, 1987, Investeringen en de financiele infrastructuur, WRR rapport V60,(Staatsuitgeverij, Den Haag).

Fama, E.F. and M.C. Jensen, 1983, Separation of ownershipfirm control, Journal of Law and Economics, 26, p. 301-325

Ferri, M.G. and W.H. Jones, 1979, Determinants of financial structure: a new methodological approach, Journal of Finance, vol. XXXIV, no. 3, June, 631-644.

Flath, D., and C.R. Knoeber, 1980, Taxes, failure costs, and optimal industry capital structure: An empirical test, Journal of Finance, vol. XXXV, no. 1, March, 99-117.

Foster, G., 1978, Financial statement analysis, (Prentice Hall Inc., Englewood Cliffs).

Freeman, C., 1982, The economics of industrial innovation (2nd edition), (Frances Pinter Publishers, London).

Haley, C., and L. Schall, 1979, The theory of financial decisions, (McGraw-Hill, New York).

Haugen, R.A. and L.W. Senbet, 1978, The insignificance of bankruptcy costs to the theory of optimal capital structure, Journal of Finance vol. XXXIII, 383-393.

Horrigan, J.O., 1968, A short history of financial ratio analysis, Accounting Review, no. 43, April, 284-294.

Hsiao, C., 1985, Benefits and limitations of panel data, Econometric Reviews, 4(1), 121-174.

Hsiao, C., 1986, Analysis of panel data, (Cambridge University Press, Cambridge)

Huisman, D., en W. Boon, 1984, Risicodragend vermogen in Nederland, (COB/SER, 's-Gravenhage).

Iperenburg, C.A. van, en D. van der Wijst, 1985, Een analyse van de financiele structuur van constructiebedrijven, Research Publikatie no. 13, Economisch Instituut voor het Midden- en Kleinbedrijf, Zoetermeer.

James, C., and R.O. Edminster, 1983, The relation between common stock returns, trading activity and market value, Journal of Finance, vol. XXXVIII, September.

Jensen, M.C., and W.H. Meckling, 1976, Theory of the firm: managerial behaviour, agency costs and ownership structure, Journal of Financial Economics, vol. 3, October, 305-360.

Jonkhart, M.J.L., 1980, Optimal capital structure and corporate debt capacity (J.H. Pasmans B.V., Den Haag).
Kemna, A.G.Z., 1987, Options in real and financial markets, (Kanters, Alblasserdam)
Kim, E.H., 1978, A mean-variance theory of optimal capital structure and corporate debt capacity, Journal of Finance, vol. XXXIII, no.1, March, 45-63.
Koedijk, C.G., en C.J.M. Kool, 1987, Venture capital in Nederland I, Rotterdamse Monetaire Studies, nummer 28, (Erasmus Universiteit, Rotterdam).
Kraus, A., and R.H. Litzenberger, 1973, A state-preference model of optimal financial leverage, Journal of Finance, September, 911-922.
Leland, H.E., and D.H. Pyle, 1977, Informational asymmetries, financial structure, and financial intermediation, Journal of Finance, vol. XXXII, no. 2, May, 371-387.
Lev, B., and S. Sunder, 1979, Methodological issues in the use of financial ratios, Journal of Accounting and Economics, no. 1, December, 187-210.
Levin, R.I., and V.R. Travis, 1987, Small company finance: what the books don't say, Harvard Business Review, Nov-Dec, 30-32.
Linden, J. van der, E.C. Waterlander en D. van der Wijst, 1988, Risicosignalering: een inventarisatie van methoden en technieken in de literatuur, Research Publikatie no. 21, Economisch Instituut voor het Midden- en Kleinbedrijf, Zoetermeer.
Marquardt, D.W., 1963, An algorithm for least squares estimation of non linear parameters, Journal of the Society for Industrial and Applied Mathematics, no. 11.
Marsh, P., 1982, The choice between equity and debt: an empirical study, Journal of Finance, vol. XXXVII, no. 1, March, 121-144.
Miller, M., 1977, Debt and taxes, Journal of Finance, vol. XXXII, no. 2, 261-275.
Modigliani, F., and M. Miller, 1958, The cost of capital, corporation finance and the theory of investment, American Economic Review, vol. 48, no. 3, June, 261-297.
Modigliani, F., and M. Miller, 1963, Taxes and the cost of capital: a correction, American Economic Review, vol. 53, no. 3, June, 433-443.
Mueller, D.C., 1972, A life cycle theory of the firm, The Journal of Industrial Economics, vol. XX, no. 3, July, 199-219.
Myers, S.C., 1977, Determinants of corporate borrowing, Journal of Financial Economics, vol. 5, November, 147-175.
Myers, S.C., 1984, The capital structure puzzle, Journal of Finance, vol. XXXIX, no.3, July, 575-592.
Myers, S.C., and G.A. Pogue, 1974, A programming approach to corporate financial management, Journal of Finance, vol. XXIX, 579-599.
Myers, S.C. and N. Majluf, 1984, Corporate financing and investment decisions when firms have information investors do not have, Journal of Financial Economics.
Neave, E.H., and J.C. Wiginton, 1981, Financial management, theory and strategies (Prentice Hall Inc., Englewood Cliffs).
Nooteboom, B., 1980, Retailing: applied analysis in the theory of the firm, (J.C. Gieben, Uithoorn).
Polak, N.J., 1950, Enige grondslagen voor de financiering der onderneming, (De Erven F.Bohn, Haarlem).
Raad voor het Midden- en Kleinbedrijf, 1979, Financiele structuur in het MKB, ('s-Gravenhage).
Remmers, L., A. Stonehill, R. Wright and Th. Beekhuisen, 1974, Industry and size as debt ratio determinants in manufacturing internationally, Financial Management, vol. III. no. 2, summer, 24-32.

Ridder, W.J. de, 1985, Overheidsstimulering van risicokapitaal; noodzaak, ideologie en marktbeinvloeding, in: W.M. van der Bergh, P.H.A.M. Verhaegen en R.E. Wessels (red.), Financiering en belegging, stand van zaken anno 1985, (Erasmus Universiteit, Rotterdam), 99–106.
Robicheck, A.A., and S.C. Myers, 1965, Optimal financing decisions, (Prentice Hall Inc., Englewood Cliffs).
Robinson, R.I., 1968, Financing the dynamic small firm, (Wadsworth Publishing Company, Belmont).
Roll, R., 1981, A possible explanation of the small firm effect, Journal of Finance, vol. 36, no. 4, September, 879–888.
Ross, S.A., 1977, The determination of financial structure: The incentive-signalling approach, Bell Journal of Economics, vol. 8, spring, 23–40.
Ross, S.A., 1976, The arbitrage theory of capital asset pricing, Journal of Economic Theory, 341-360.
Rubinstein, M., 1973, A mean-variance synthesis of corporate financial theory, Journal of Finance, March, 167–181.
Schuit, J.W.R., en M.C. van Gelder, 1987, Venture capital in Nederland, een onderzoek, Rapport 8715/F, Centrum voor Bedrijfseconomisch Onderzoek, Erasmus Universiteit, Rotterdam.
Scott, D.F., and J.D. Martin, 1975, Industry influence on financial structure, Financial Management, vol. IV, no. 1, spring, 67–73.
Scott, J.H., 1976, A theory of optimal capital structure, Bell Journal of Economics, 7, 33–54.
Silberberg, E., 1981, The structure of economics: A mathematical analysis, (McGraw-Hill, New York).
Sprenger, K-A., P. von Hinten und J. Steiner (Bearbeiter), 1982, Finanzierungssituation und Finanzierungsverhalten mittelstaendischer Betriebe, (Verlag Otto Schwartz & Co., Goettingen).
Spronk. J., and D. van der Wijst, 1988, Determinants of financial structure in retailing, Research Paper 8801, Economisch Instituut voor het Midden- en Kleinbedrijf, Zoetermeer.
Stiglitz, J.E., 1969, A re-examination of the Modigliani-Miller theorem, American Economic Review, vol. 59, no. 5, December, 784–793.
Stiglitz, J.E., 1974, On the irrelevance of corporate financial policy, American Economic Review, vol. 64, no. 6, December, 851–866.
Struck, P.L., and L. Mandell, 1983, The effect of bank deregulation on small business: a note, Journal of Finance, vol. XXXVIII, no. 3, June, 1025–1031.
Tamari, M., 1980, The financial structure of the small firm – an international comparison of corporate accounts in the U.S.A., France, U.K., Israel and Japan, American Journal of Small Business, vol. IV, no. 4, spring (April–June), 20–34.
Taub, A.J., 1975, Determinants of the firm's capital structure, The Review of Economics and Statistics, vol. LVII, no. 4, November, 410–416.
Tempelaar, F.M., 1985, Een kritische evaluatie van de theorie van de optimale vermogensstructuur, in: W.M. van der Bergh, P.H.A.M. Verhaegen en R.E. Wessels (red.), Financiering en belegging, stand van zaken anno 1985, (Erasmus Universiteit, Rotterdam), 181–199.
The state of small business: a report of the president, 1984, (United States Government Printing Office, Washington).
The state of small business: a report of the president, 1986, (United States Government Printing Office, Washington).
The state of small business: a report of the president, 1987, (United States Government Printing Office, Washington).
Thurik, A.R., 1984, Quantitative analysis of retail productivity, (W.D. Meinema B.V., Delft).
Thurik, A.R., and D. van der Wijst, 1984, Part-time labor in retailing, Journal of Retailing, vol. 60, no. 3, fall, 62–80.

Titman, S., 1984, The effect of capital structure on a firm's liquidation decision, Journal of Financial Economics, vol. 13, no. 1, March, 137-151.
Titman, S., 1983, Determinants of capital structure: an empirical analysis, Graduate School of Management (University of California, Los Angeles).
Titman, S., and R.E. Wessels, 1985, Determinants of capital structure choice, (working paper), University of California, Los Angeles.
Titman, S., and R. Wessels, 1985, The determinants of capital structure, Graduate School of Management, (University of California, Los Angeles).
Toy, N., A. Stonehill, L. Remmers, R. Wright and Th. Beekhuisen, 1974, A comparative international study of growth profitability and risk as determinants of corporate debt ratios in the manufacturing sector, Journal of Financial and Quantitative Analysis, vol. IX, no. 5, November, 875-886.
Viscione, J.A, 1986, How long should you borrow short term?, Harvard Business Review, March-April, 20-24.
Walker, E.W., and J.W. Petty, 1978a, Financial management of the small firm, (Prentice Hall Inc., Englewood Cliffs).
Walker, E.W., and J.W. Petty, 1978b, Financial differences between large and small firms, Financial Management, vol. VII, no. 4, winter, 61-68.
Warner, J., 1977, Bankruptcy costs : some evidence, Journal of Finance, May, 337-347.
Welsh, J.H., and J.F. White, 1981, A small business is not a little big business, Harvard Business Review, July-August, 18-32.
Weston, J.F., and E.F. Brigham, 1972, Managerial finance, (Holt, Rinehart and Winston, London).
Wijst, D. van der, 1985, De ontwikkelingen in vermogensverhoudingen, Pecunia, jrg. 1, no. 1, 5-9.
Wijst, D. van der, 1986, Corporate debt policy and valuation of corporate claims, Research Paper 8601, Economisch Instituut voor het Midden- en Kleinbedrijf, Zoetermeer.
Williamson, O.E., 1975, Markets and hierarchies: analysis and antitrust implications, (The Free Press, New York).
Woodcock, C., 1986, The financial and capital environment of the small firm, in: James Curran, John Stanworth and David Watkins (ed.), The survival of the small firm 1, the economics of survival and entrepreneurship, (Gower, London), 39-52.
Wytzes, H.C., 1975, Ondernemingsfinanciering, theorie en politiek, (H.E. Stenfert Kroese, Leiden).

AUTHOR INDEX

SUBJECT INDEX